AN INTRODUCTION TO UNCERTAINTY IN MEASUREMENT USING THE GUM (GUIDE TO THE EXPRESSION OF UNCERTAINTY IN MEASUREMENT)

Measurement shapes scientific theories, characterises improvements in manufacturing processes and promotes efficient commerce. Inherent in measurement is uncertainty, and students in science and engineering need to identify and quantify uncertainties in the measurements they make. This book introduces measurement and uncertainty to second- and third-year students of science and engineering. Its approach relies on the internationally recognised and recommended guidelines for calculating and expressing uncertainty (known by the acronym GUM). The statistics underpinning the methods are considered and worked examples and exercises are spread throughout the text. Detailed case studies based on typical undergraduate experiments are included to reinforce the principles described in the book. This book is also useful to professionals in industry who are expected to know the contemporary methods in this increasingly important area. Additional online resources are available to support the book at www.cambridge.org/9780521605793/.

LES KIRKUP has more than 25 years' experience of working in tertiary-education institutions. He held academic positions in England and Scotland before moving to Australia in 1990. He is currently an associate professor in the Faculty of Science at the University of Technology, Sydney. He holds a B.Sc. in Physics (Sheffield), an M.Sc. in Solid State Physics (London) and a Ph.D. (Paisley). He is a member of the Institute of Physics, the Australian Institute of Physics and the Metrology Society of Australia. He is the author of three books, including *Data Analysis with Excel* (2002, Cambridge University Press). A dedicated lecturer, many of Kirkup's educational developments have focussed on improving the laboratory experience of undergraduates drawn from the physical sciences and engineering. He is an active researcher and he is currently studying band broadening in high-performance liquid chromatography. He has published in a broad range of journals, including *Computer Applications in the Biosciences*, *Computers in Physics*, *Review of Scientific Instruments*, *Medical and Biological Engineering and Computing*, *Physics Education*, *European Journal of Physics*, *American Journal of Physics*, *Physiological Measurement*, *Measurement Science and Technology* and *Journal of Chromatography A*.

BOB FRENKEL has an M.Sc. in Physics from the University of Sydney and an M.Eng.Sc. from the University of New South Wales. He is a senior experimental scientist in electrical standards at the Australian National Measurement Institute and is responsible for the maintenance and development of the Australian national standard of dc voltage. His publications in this and related fields include articles in *Metrologia, Measurement Science and Technology, Journal of Physics E: Scientific Instruments, Transactions of the IEEE (Instrumentation and Measurement), Journal of Electrical and Electronics Engineering* (Australia), *Transactions of the Instrument Society of America* and *Proceedings of the Metrology Society of Australia,* of which he is a founding member. He is an electrical assessor with the National Association of Testing Authorities of Australia. His other professional interest is statistics in metrology, an area of rapidly growing importance since the publication and dissemination in the mid 1990s of the International Standardisation Organisation's *Guide to the Expression of Uncertainty in Measurement (GUM).* As a member of the National Measurement Institute's Uncertainty Panel he is a consultant on statistical issues and has taken part in educational seminars on the GUM.

AN INTRODUCTION TO UNCERTAINTY IN MEASUREMENT USING THE GUM (GUIDE TO THE EXPRESSION OF UNCERTAINTY IN MEASUREMENT)

L. KIRKUP AND R. B. FRENKEL

CAMBRIDGE
UNIVERSITY PRESS

CAMBRIDGE
UNIVERSITY PRESS

University Printing House, Cambridge CB2 8BS, United Kingdom

Cambridge University Press is part of the University of Cambridge.

It furthers the University's mission by disseminating knowledge in the pursuit of
education, learning and research at the highest international levels of excellence.

www.cambridge.org
Information on this title: www.cambridge.org/9780521605793

First published 2006

A catalogue record for this publication is available from the British Library

ISBN 978-0-521-84428-4 Hardback
ISBN 978-0-521-60579-3 Paperback

Cambridge University Press has no responsibility for the persistence or accuracy of
URLs for external or third-party internet websites referred to in this publication,
and does not guarantee that any content on such websites is, or will remain, accurate
or appropriate.

Cover illustration: Motorbike stunt in Sydney Harbour.
Photograph courtesy of Geoff Ambler.

To our families

Contents

And we know today – how little we know. There is never an observation made but a hundred observations are missed in the making of it; there is never a measurement but some impish truth mocks us and gets away from us in the margin of error.

H. G. Wells, *Men Like Gods* (1923)

Exact transcription from *Men Like Gods*, by H. G. Wells,
as published by The Macmillan Company, New York, 1923.

Preface

In writing this book, we address several groups of readers who require an understanding of measurement, and of uncertainty in measurement, in science and technology.

Undergraduates in science, for example, should have texts that set out the concepts and terminology of measurement in a clear and consistent manner. At present, students often encounter texts that are mutually inconsistent in several aspects. For example, some texts use the terms *error* and *uncertainty* interchangeably, whilst others assign them distinctly different meanings. Such inconsistency is liable to confuse students, who are consequently unsure about how to interpret and communicate the results of their measurements.

Until recently, a similar lack of consistency affected those whose primary occupation includes measurement, the evaluation of uncertainty in measurement, instrument and artefact calibration and the maintenance of standards of measurement – that is, professional metrologists. International trade, for example, requires mutual agreement among nations on what uncertainty is, how it is calculated and how it should be communicated; for a global economy to work efficiently, lack of such agreement cannot be tolerated. In the mid 1990s, international bodies, charged with the definition, maintenance and development of technical standards and standards of measurement in a variety of fields, published and disseminated the *Guide to the Expression of Uncertainty in Measurement* – the 'GUM'. These bodies included the Bureau International des Poids et Mesures (BIPM) or International Bureau of Weights and Measures, the International Standardisation Organisation (ISO) and the International Electrotechnical Commission (IEC). The GUM is being adopted worldwide by organisations representing a diversity of disciplines, such as calibration and testing laboratories in the physical and engineering sciences, in chemical and biochemical analytic work and related specialised areas of medical testing, in the certification of reference materials and, at the highest metrological level, in national measurement institutes.

Despite its prominence in all fields of measurement, the GUM is (in 2005) largely unknown amongst university and college academics. One of our goals in writing this book is to introduce the GUM and its essential statistical background to an undergraduate audience. We believe that adopting the methods described in the GUM at undergraduate level will confer improved clarity and consistency on the teaching and learning of errors and uncertainty, and on their expression. As use of the GUM grows in industrial and commercial laboratories, new generations of graduating students will require a working knowledge of its methods and vocabulary as well as of the statistical principles that underpin them. In this book we have attempted to anticipate and address these needs.

We include introductory material in the early chapters, the level of which is consistent with first-year university courses. However, the book as a whole is likely to be of greater benefit to second-year students who have already had some exposure to laboratory work as well as a first-year course in calculus and some basic statistics. When dealing with statistical relationships, we have not attempted the rigour normally found in mathematical statistical texts, but have preferred to introduce them in an intuitively plausible way, often by means of figures and graphs.

We have made some use of Monte Carlo simulation (MCS) in the text. One reason is that the GUM, which advocates as a standard practice the law of propagation of uncertainties involving first-order derivatives of the inputs, nonetheless recognises the need for 'other analytical or numerical methods' (when a complicated relationship exists between a measurand and its inputs). One such method is MCS, and therefore some exposure to MCS is desirable. Another important reason, in an educational context, is that MCS can make a statistical process, summarised by a theoretical equation, 'transparent' to the reader in a way that a standard theoretical approach does not. MCS bears much the same relationship to theoretical statistics as experimental physics does to theoretical physics, and can be a valuable and accessible teaching tool, since all it requires is a personal computer, random-number-generating software and some programming or spreadsheet experience.

As part of the text we have also introduced and described in some detail particular undergraduate experiments. The 'real' data from these experiments are analysed using the methods described in this book. These experiments might be suitable for adaptation to courses that deal with measurement and uncertainty. Anticipating this, we have included some suggestions on how the experiments might be developed or enhanced.

The need for an exposition of the GUM extends beyond the universities. An equally important group towards whom this book is directed consists of professionals who are not necessarily involved in making measurements or assessing uncertainties on a day-to-day basis, but who must nevertheless be familiar with contemporary international guidelines relating to measurement and uncertainty.

Specialist publications that deal with uncertainty and the GUM often assume that the reader is, or will become, a practising metrologist, and are therefore written at an advanced level. We hope that this book, with its combination of general principles and specific examples, will aid those readers who wish to know something of current guidelines, but who are less inclined to consult such publications. We have, however, included in several places some slightly more advanced material; the reason is that there occur, not uncommonly, situations where simple formulas as presented in introductory texts become invalid, and to understand why, this more advanced material is needed.

The calculation and expression of uncertainty constitute only one aspect of measurement. The subject also includes the detection, description, analysis and minimisation of errors. The minimisation of errors, moreover, can be achieved in aesthetically pleasing ways that are often pioneered at national measurement institutes. We have, therefore, included some discussion of these topics. Since in any attempt at accurate measurement there is likely to be a variety of potential sources of error, a broad familiarity with several branches of science is desirable in metrology. The best metrologists, in fact, tend to be scientific 'all-rounders'. We have given some examples of the need for this professional versatility, which contributes to the fascination and challenge of measurement in science and technology.

One of the great rewards of writing a book is the amount learned by the author or authors along the way. 'To teach is to learn' is ancient wisdom. This is certainly true for us, and we acknowledge many people who have helped us clarify our thinking. We are grateful for the assistance we have received in our attempt to bridge the gap between a textbook written by an academic for an academic audience and a specialist text written for practising metrologists by a professional in metrology.

We warmly acknowledge the contribution of our colleagues and peers who have suggested examples, problems and topics for inclusion in the text. We thank, from the National Measurement Institute of Australia (NMIA) in Sydney, Errol Atkinson, Mark Ballico, Robin Bentley, Noel Bignell, Nick Brown, Ilya Budovsky, Henry Chen, Jim Gardner, Åsa Jämting, John Peters, Steve Quigg, Brian Ricketts and Greig Small; from the University of Technology, Sydney (UTS), Nick Armstrong, Sherran Evans, Matthew Foot, Jim Franklin, Suzanne Hogg, Walter Kalceff, Geoff McCredie and Greg Skilbeck; from the Stunt Agency, Jennifer Fenton and, from Cambridge University Press, Simon Capelin and Vince Higgs. We also thank Alan Johnston for advice on cover design, and Avril Wynne for her daring flight. We also gratefully acknowledge the love, support and forbearance of our families throughout the writing of this book.

Les Kirkup and Bob Frenkel, May 2005

1

The importance of uncertainty in science and technology

We live with uncertainty every day. Will the weather be fine for a barbecue at the weekend? What is the risk to our health posed by a particular item of diet or environmental pollutant? Have we invested our money wisely?

It is understandable that we would like to be able to eliminate, or at least reduce, uncertainty. If we can reduce it significantly, we become more confident that a desirable event will happen, or that an undesirable event will not. To this end we seek out accredited professionals, such as weather forecasters, medical researchers and financial advisers.

However, in science and technology uncertainty has a narrower meaning, created by the need for accurate measurement. Accurate measurement, which implies the existence of *standards* of measurement, and the evaluation of uncertainties in a measurement process are essential to all areas of science and technology. The branch of science concerned with maintaining and increasing the accuracy of measurement, in any field, is known as *metrology*.[1] It includes the identification, analysis and minimisation of errors, and the calculation and expression of the resulting uncertainties.

Whether or not a measurement is regarded as 'accurate' depends on the context. Supermarket scales used for weighing fruit or vegetables need not be better than 1% accurate. By contrast, a state-of-the-art laboratory balance is able to determine the value of an unknown mass of nominal value one kilogram[2] to better than one part in ten million. These figures, 1% in one case and one part in ten million in the other, are numerical measures of the degree of accuracy: low in the first case and high in the second, but each of them fit for its particular purpose. Evidently, accuracy and

---

[1] This word derives from the Greek 'to measure'. It should not be confused with *meteorology*, the study of climate and weather. The need for accurate measurement, and for standards of length, weight and volume (for example), was recognised in many ancient societies with relatively primitive technology and hardly any 'science' in the modern sense.

[2] A 'nominal' value is the ideal or desired value of a particular quantity. Thus the nominal value of the mass of an object might be 1 kilogram, implying that its accurately measured value is close to 1 kilogram.

uncertainty are inversely related: high accuracy implies low uncertainty; and low accuracy implies high uncertainty.

When we say that the result of a measurement has an associated uncertainty, what, exactly, are we uncertain *about*? To begin to answer this question, we acknowledge that the result of a measurement is usually a *number* expressed as a multiple of a *unit* of measurement. As in the example of the laboratory balance above, we should refer to a number that results from a measurement as a *value*. For example, a person's mass may have a value of 73 kilograms, meaning that the mass is 73 units, where each unit is one kilogram. Similarly, the temperature of coffee in a cup may be 45 degrees Celsius, the length of a brick 231 millimetres, the speed of a car 60 kilometres per hour, and so on. The value that expresses the given quantity therefore depends on the unit. The same speed of the car, for example, could be expressed as 17 metres per second. There are cases where the value is independent of the unit. This happens when a quantity is defined as a ratio of two other quantities, both of which can be measured in terms of the same unit. The units then 'cancel out'. For example, the coefficient of static friction, μ_s, is defined as the ratio of two forces and therefore μ_s is a *dimensionless* number; for glass on glass, $\mu_s \simeq 0.94$.

A measurement whose result is characterised by a value holds more information than a measurement whose result is not characterised in this way. In the latter case we might hesitate to call the result a 'measurement'; it would be more in the nature of an opinion, judgment or assessment. In fact, this is how we tend to function in everyday life. When parking a car in a busy street, the driver estimates the available space in most – though not all – cases quite adequately without a rule or tape-measure. We may think a person handsome or beautiful, but it would be rash to attempt seriously to attach a numerical value to this. (If we drop the word 'seriously', then it is possible. A 'millihelen' may be defined as the amount of beauty required to launch exactly one ship![3])

The information-rich use of a value to characterise the result of a measurement comes at a price. We should also consider – particularly in pure and applied science, in medicine and in engineering – how 'uncertain' that value is. Is the length of the brick 231 millimetres, or more like 229 millimetres? What is the most appropriate instrument for measuring the length of the brick, and how can we be sure of the accuracy of the instrument? How, in any case, do we define the 'length' of a brick, which may have rough or uneven edges or sides? How much 'leeway' can we afford to allow for the length of a brick, before we must discard it as unusable?

This book considers measurement, uncertainty in measurement and, in particular, how uncertainty in measurement may be quantified and expressed. International

---

[3] This refers to a story from ancient Greece, as recounted by Homer in the *Iliad* around the eighth century BC. The beautiful Helen of Sparta, in Greece, had been taken to Troy (in what is now Turkey), and that started the ten-year Trojan War. The Greeks launched a fleet of one thousand ships to reclaim her.

guidelines exist to assist in these matters. The guidelines are described in the *Guide to the Expression of Uncertainty in Measurement*, published by the International Standardisation Organisation (corrected and reprinted version in 1995), abbreviated as 'the GUM'. Before discussing and illustrating these guidelines in detail, we highlight the importance of measurement and uncertainty by considering some examples.

1.1 Measurement matters

Just how important are measurement and uncertainty? Careful measurement with properly identified and quantified uncertainties could lead to a new discovery and international recognition for the scientist or scientific team that made the discovery. To the engineer it may lead to improved safety margins in complex systems such as those found on the space shuttle, and to the police it could contribute to the successful prosecution of a driver who exceeds the speed limit in a motor vehicle. In biochemical metrology, accurate measurement is needed for the reliable estimation of (for example) trace levels of food contaminants such as mercury in fish. In medical metrology, high accuracy in blood-pressure measurements reduces the risk of misdiagnosis. We now give some examples of advances in measurement accuracy. At the end of this chapter we indicate where further information on these topics may be found. We use the SI (Système International)[4] units of measurement, which include the metre (m) for distance, the kilogram (kg) for mass and the second (s) for time.

1.1.1 Measurements of the fundamental constants of physics

Theories of the physical world incorporate fundamental constants such as the speed of light, c, the Planck constant, h, the fine-structure constant, α, and the gravitational constant,[5] G. As far as we know, these are true constants: they do not change with time or location and have the same values on Earth as anywhere else in the Universe. In many cases their numerical values are accurately known, and in a few cases the constants have been exactly defined. For example, the speed of light, c, in a vacuum is defined as $c = 299\,792\,458 \, \text{m} \cdot \text{s}^{-1}$. The Planck constant, h, which is the ratio of the energy of a photon of radiation to its frequency, is accurately known: $h = 6.626\,069 \times 10^{-34} \, \text{J} \cdot \text{s}$ (joule-second) with an uncertainty of less than one part in a million.

---

[4] The French acronym is universally used in recognition of the central role played by France, during the late eighteenth century and later, in introducing and establishing the uniform system of units of measurement that came to be known generally as the 'metric' system and that later evolved into the SI.

[5] 'Big G' is not to be confused with g, 'little g', which is the acceleration due to gravity near the Earth's surface and varies with location.

The gravitational constant, G, appears in the equation that describes the inverse-square law of gravitation discovered by Isaac Newton in the seventeenth century: $F = Gm_1m_2/r^2$, where F is the gravitational force of attraction between two masses m_1 and m_2 a distance r apart. To calculate the force using this equation, we must know the value of G. With the practically available masses in a laboratory this force is tiny because G is very small: about $6.68 \times 10^{-11} \text{m}^3 \cdot \text{kg}^{-1} \cdot \text{s}^{-2}$. For example, two uniform spherical bodies, each of mass 200 kg, whose centres are separated by 1 m (these could be two solid steel spheres each of approximate diameter 36 cm) would attract each other with a gravitational force of about 2.7×10^{-6} N. This is roughly one-tenth the weight of a small ant (mass $\simeq 3$ mg).

We have a healthy respect for the Earth's gravitational force, but this is largely due to the enormous mass of the Earth, about 6×10^{24} kg (this mass has to be inferred from a known value of G). In measuring G, the tiny gravitational forces that exist between bodies in a laboratory make an accurate measurement of G very difficult. These tiny forces must somehow be measured against a background of competing gravitational forces, including the much larger ordinary gravity due to the Earth as well as the gravity exerted by the mass of the scientist doing the experiment! At the time of writing (2005), the accepted fractional uncertainty in G is about one part in ten thousand. This is much larger than the fractional uncertainty with which other fundamental constants are known. Previous attempts to measure G made in the 1990s yielded results that were mutually discrepant by several parts per thousand, even though much smaller uncertainties were claimed for some of the individual results.[6] Experiments to measure G accurately are evidently beset by subtle *systematic* errors (systematic errors will be discussed later in this book).

When G or any other particular quantity is measured, it is important to know the uncertainty of the measurement. If two values are obtained for the same particular quantity, and these values differ by significantly more than the uncertainty attached to each value, then we know that 'something is wrong': the quantity has perhaps undergone some change in the interval between the two measurements, or systematic errors have not been properly accounted for. The latter interpretation is evidently the more likely one with respect to the determination of G.

Painstaking measurements of G, and of other fundamental constants, yield new insights into our physical world. In applied physics and engineering, seeking reasons for discrepancies often leads to better understanding of materials or of laboratory techniques. In the case of G, where several experiments have been based on the twisting of a strip of metal (a 'torsion strip') in response to the gravitational field of nearby masses, it has been found that such torsion strips are not perfectly elastic (that is, the amount of twist is not exactly proportional to the torque), and the

---

[6] Figure 4.2 in chapter 4 illustrates this.

amount of this so-called 'anelasticity' is significant. This finding is a contribution to knowledge in its own right. In theoretical physics, high-accuracy measurements of G will eventually contribute usefully to current speculation as to whether there may be some small but detectable violation of the inverse-square law over laboratory distances and even at the sub-millimetre level. Such violations would have profound implications for our understanding of the Universe. It is only through careful measurement and realistic estimates of uncertainty that we can have confidence in any conclusions drawn from results of studies designed to establish a value for G.

1.1.2 Careful measurements reveal a new element

At the end of the nineteenth century, Lord Rayleigh showed the benefits that accrue from close scrutiny of the results of measurements that appear at first glance to be consistent and to contain nothing very surprising. Rayleigh used two methods to measure the density of nitrogen.[7] In one method, the nitrogen was obtained wholly from the atmosphere, by passing air over red-hot copper that removed all the oxygen. In the other method, the nitrogen was obtained by bubbling air through ammonia and then passing the air–ammonia mixture through a red-hot copper tube. This also removed the oxygen (which combined with hydrogen from the ammonia to form water), but partly 'contaminated' the nitrogen from the air with nitrogen from the ammonia itself. The nitrogen obtained by the second method (the 'chemical' method) was about 0.1% less dense than that given by the first method (the 'atmospheric' method). Despite the close agreement, Rayleigh was uncomfortable with the 0.1% discrepancy and resisted his instinct to find ways to downplay or ignore the difference. Instead, he undertook a detailed study in which he tried to exaggerate the difference by varying the experimental conditions. He replaced the air in the chemical method by pure oxygen, so that all the collected nitrogen originated from the ammonia. This modified chemical method now provided nitrogen that was 0.5% less dense than that obtained by the atmospheric method. Thus Rayleigh had strong evidence that nitrogen derived from the atmosphere had a (very slightly) greater density than nitrogen derived from 'chemical' sources (for example, ammonia).

The inescapable conclusion of Rayleigh's careful measurements was that his atmosphere-derived 'nitrogen' was in fact nitrogen mixed with another gas. The gas that Rayleigh had discovered was argon, a new element, and for this discovery Rayleigh was awarded the Nobel prize in physics in 1904. While 78% of the atmosphere is nitrogen, only about 1.2% is argon, but argon is denser than nitrogen

---

[7] Rayleigh measured a mass of nitrogen. Since this was done at a standard temperature and pressure, the volume of nitrogen was fixed, so effectively its density was measured.

by a factor of about 1.4. So atmosphere-derived nitrogen, containing unidentified argon, appeared to be denser than chemical-derived nitrogen.

Rayleigh's original measurements of the collected mass of nitrogen were made with an uncertainty of about 0.03% or less. A larger uncertainty might easily have obscured the small (0.1%) systematic discrepancy that compelled him to pursue the matter further. This story illustrates the need for accurate measurement, the benefit gained by measuring a quantity in more than one way and the importance of explaining any discrepancy thereby revealed.

Since Rayleigh's time, experimental methods and instruments have advanced significantly so that, for example, instruments under computer control can gather vast amounts of data in a very short time. With respect to measurement and uncertainty, this brings its own challenges.

1.1.3 Treat unexpected data with caution

In 1985 scientists doing atmospheric research in Antarctica announced that the ozone layer over the South Pole was being depleted at quite a dramatic rate. Their conclusion was based on ground measurements of ultraviolet radiation from the Sun that was absorbed by the atmosphere. For several years prior to this, other scientists had been 'looking down' on the ozone layer using satellites, though they had reported no change in the depth of the layer. A contributory factor to the inconsistency between the ground-based and satellite-based data could be traced to the processing of the satellite data. Natural variation in values of the thickness of the ozone layer was well known. Therefore it appeared reasonable, when processing the satellite-based data, to discard 'outliers' – that is, data that appeared not to conform with that natural variation. The problem with this approach was that, if the 'natural' variation were *itself* changing, one risked discarding the very data that would reveal such a change. When the satellite data were reanalysed with the outliers included, the conclusion of the Antarctica scientists was supported. The effect of this prominent work was to fuel international debate among scientists, industry and governments on the causes, consequences, extent and treatment of ozone depletion in the atmosphere.

Quantifying ozone depletion by investigating the absorption of ultraviolet radiation by the Earth's atmosphere is an example of the application of optically based measurement. Optically based methods of measurement are widely used, and many rely on that most versatile of devices, the laser.

1.1.4 The laser and law-enforcement

The laser ('light amplification by stimulated emission of radiation'), invented and developed in the early 1960s, offers very high accuracy in length measurement in research and industry. Laser interferometry is a standard technique used in industry to measure length to sub-micrometre precision. This is made possible by the monochromatic ('single-colour') nature of laser light, implying a single wavelength and therefore a natural 'unit of length'. The red light from an iodine-stabilised helium–neon laser has a wavelength of 632.991 212 58 nm (nanometres or 10^{-9} m), with an uncertainty of the order of a few parts in 10^{11}. A measurement of length can, therefore, be 'reduced' to counting wavelengths: more precisely, the counting of interference fringes that result from the interference of the beam of laser light with a similar reference beam.

Applications of lasers even extend to law-enforcement. The speed of a vehicle can be established by aiming a narrow beam of pulsed infra-red radiation emitted by an instrument containing a laser (the 'speed-gun') at the body of the moving vehicle. The pulses are emitted at an accurately known rate of the order of 100 pulses every second. The radiation is reflected by the body and returns to the instrument. If the vehicle is moving towards the speed-gun, the interval between successive reflected pulses is less than the interval between successive transmitted pulses. This difference is small, of the order of nanoseconds (or billionths of a second), but can be accurately measured. This difference and the known value of the speed of light enable the speed of the vehicle to be determined. Speeds recorded well in excess of the speed limit can lead to instant licence disqualification in some countries, and an appearance in court. Identifying and understanding the complications that may affect the value measured for the vehicle speed is the starting point for estimating the uncertainty of the measurement of speed. Such complications include the exact angle of the speed-gun relative to the direction of the vehicle, interfering effects of bright light sources and whether the speed-gun has been accurately calibrated and is not significantly affected by variations in ambient temperature. It is only when the uncertainty in the speed is known that it is possible to decide whether a vehicle is very likely to be exceeding the speed limit.

1.1.5 The Global Positioning System (GPS)

A GPS receiver can determine its position on the Earth with an uncertainty of less than 10 metres. This is made possible by atomic clocks carried on satellites orbiting the Earth with an approximate half-day period and at a distance of about 20 000 kilometres. The atomic clocks are stable to about one part in 10^{13} (equivalent to gaining or losing one second in about 300 000 years). Atomic clocks of this degree of stability evolved from research by Isador Rabi and others in the 1930s and later on the natural resonance frequencies of atoms. The receiver contains its own clock (which can be less stable) and, by comparing its own clock-time with the transmitted satellite clock-times, the receiver can calculate its own position. The comparison of clock-times must take into account the first-order Doppler shift, of about one part in 10^5 in the case of the GPS, of the frequency of a clock moving towards or away from a fixed clock.[8] A further requirement for the accuracy of the GPS is the relativity theory of Albert Einstein. Two of the relativistic effects that must be taken into account are the slowing (time-dilation) of satellite clocks moving transversely relative to fixed clocks (this is also known as the second-order Doppler shift) and the speeding up of clocks far from the Earth's surface due to the weaker gravitational field. These two effects act in opposition and have magnitudes of about one part in 10^{10} and five parts in 10^{10}, respectively. So two major branches of theoretical physics have made possible timekeeping metrology of extremely high accuracy and have revealed subtle properties of time and space. As a result, inexpensive devices that accurately determine the location of aircraft, ships and ground vehicles, and help with the safety of explorers and trekkers, are now available.

1.1.6 National measurement institutes, international metrology and services to industry

It is obvious that industrial products must perform reliably. This implies something that is perhaps not so obvious: the relevant physical properties of their components must be certified against local and, ultimately, international standards of measurement. Such standards are very precisely and meticulously manufactured objects, for example steel rules and tape-measures, standard weights, standard resistors and standard lamps. If component A of a motor-vehicle (for example) must fit or be compatible with component B, this certification will ensure that, if A is made in country X and B in country Y, A will fit B in country Z where the motor-vehicle is assembled. International certification depends on the existence of standards of measurement in every field of science and technology. Research into, and the development and maintenance of, standards of measurement at the highest possible level of accuracy are the function and responsibility of a country's national measurement institute (NMI).

For a physical property of a component to be certified, it must be compared with or *calibrated against* the relevant standard. If the component is (for example) a 1000-Ω resistor, its resistance will be compared with a local 1000-Ω *standard* resistance, which may, however, be of relatively low accuracy. This standard, in turn, must be calibrated against a higher-accuracy standard, generally maintained by industrial calibration laboratories, and so on until the top of the comparison chain is reached. This would normally be the national standard of resistance maintained

---

[8] This first-order Doppler effect is familiar to us in its acoustic analogue as the raised pitch of the sound made by an approaching object, and the lowered pitch when it recedes.

by an NMI, and would itself be validated by frequent international comparisons by various NMIs of such national standards (or of very stable and highly accurate standards directly *traceable* to such national standards). The degree to which the participating NMIs' standards 'agree with one another' or, more formally, have the essential property of 'mutual equivalence' is a decision made by the BIPM (Bureau International des Poids et Mesures, or International Bureau of Weights and Measures, in Paris). Such international comparisons are a routine feature of international metrology, and serve to maintain the reliability and underpin the quality control of a huge variety of industrial products in day-to-day trade and commerce.[9]

Progress in metrology – namely, permanently improved standards and reduced uncertainties – is usually made by an NMI, although occasionally by other institutions. This happens through a major change in method inspired by a novel application of existing knowledge, or by use of an advance in physics or other science. There are many such examples; two will be described here, while other cases will be mentioned later in the book.

1.1.6.1 Standards of electrical resistance and capacitance

The history of the standard of resistance provides a good example of the kind of research, often in seemingly unrelated areas, that informs progress in metrology. For about the first half of the twentieth century the 'international ohm' was defined as the resistance of a specified length and volume of mercury at a specified temperature. The complicating factors here are the inevitable uncertainties in the measurements of the length, volume and temperature of the mercury, and uncertainty regarding its purity.

Another metrological route towards a standard of resistance could be found if a standard of capacitance could be defined. These are two quite different electrical quantities measured in different units, but there is a simple relationship between them. Unfortunately, a capacitance, C, is normally physically constructed as two metal plates separated by an insulating gap (assumed here to be a vacuum), and so is calculated using an expression of the form $C = \epsilon_0 A/d$, where ϵ_0 is the constant permittivity of free space (or vacuum),[10] A is the area of the capacitor plates and d is their separation (figure 1.1(a)). The uncertainty in C will now result from the considerable uncertainties in the measurements of A and d.

---

[9] To maintain standards (of performance as well as measurement) private and government laboratories, and NMIs themselves, undergo regular review by assessors. Successful review is followed by accreditation of the laboratory for its particular area of expertise. The NMIs are accredited through international comparisons and by means of peer-review by visiting experts from other NMIs.

[10] The value of ϵ_0, a natural constant, in SI units to eight significant figures is $8.854\,187\,8 \times 10^{-12}$ F·m^{-1}. Ordinary capacitors as used routinely in electronics have insulating material (a 'dielectric'), rather than a vacuum, between the plates. The effective permittivity is then larger than ϵ_0.

It is important to note that ϵ_0 in the expression $C = \epsilon_0 A/d$ has units farad per metre. The product $\epsilon_0 A/d$ consequently has units (farad per metre) $\times$ metre^2/metre, or farad, equal to the units of C. A major advance in capacitance and resistance metrology would therefore result if a geometry could be found in which C was given simply as ϵ_0 multiplied by a distance, since this product would also have the units of a capacitance: farad per metre $\times$ metre gives farad. In effect, the nuisance of having to measure an area and a length would be replaced by the convenience of measuring only a length.

Using both mathematical analysis (starting from Maxwell's equations of electrostatics) and experimental verification, such a geometry was found in 1956 by A. M. Thompson and D. G. Lampard of the National Standards Laboratory of Australia (now known as the National Measurement Institute). This discovery became known as the Thompson–Lampard Theorem of Electrostatics. The most common practical realisation of this theorem is shown in figure 1.1(b) and has come to be known as the 'calculable capacitor'. Four identical circular cylinders A, B, C and D, each centred at the corner of a square, are enclosed within a circular earthed shield E and are separated from one another and from the shield by narrow insulating gaps. There are two earthed central bars. Only one of these (F) is shown, and F is movable perpendicular to the plane of the diagram. If F is moved a distance d, it can be shown that the resulting change, C, in capacitance ('cross-capacitance') between A and C (with B and D earthed) or between B and D (with A and C earthed) is given by $C = \epsilon_0[(\log 2)/\pi]d$. This is a small change, approximately 2 pF per metre. The distance d can be very accurately measured using laser interferometry.

We therefore note the crucial geometry-independent property of the calculable capacitor: the capacitance depends only on d, not on (for example) the diameters of the cylinders. Figure 1.1(b) could be scaled up or down in the plane of the diagram by any factor, and C would still be given as stated above.[11] In electrical metrology, geometry-independence is a prized attribute of any measurement that strives towards the highest accuracy.

Standard resistors of nominal value 1 Ω can be calibrated against the calculable capacitance C by means of well-established procedures. The calculable capacitor has, therefore, provided a realisable 'absolute' ohm, a primary standard much superior to the 'international' ohm mentioned previously. The uncertainty of the resistances is of the order of a few parts in a hundred million, and these form the primary standards for disseminating the practical unit of resistance throughout the research and industrial communities.

---

[11] Figure 1.1(b) is a particular case of a more general configuration involving four surfaces separated by narrow gaps. For this general case, C is given by a formula that still involves only a single distance measurement d and that reduces to $C = \epsilon_0[(\log 2)/\pi]d$ for figure 1.1(b).

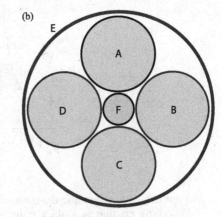

(b)

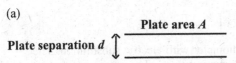

(a)

Plate area *A*

Plate separation *d*

Figure 1.1. (a) Capacitance C given by $C = \epsilon_0 A/d$. (b) The calculable capacitor. Four identical cylinders A, B, C and D are centred at the corners of a square with narrow gaps between neighbours. E is an earthed shield and F is a central bar movable perpendicular to the plane of the diagram. If F is moved by a distance d, the change in cross-capacitance, C, (between A and C, or B and D) is given by $C = \epsilon_0[(\log 2)/\pi]d$.

1.1.6.2 *The cryogenic radiometer*

Many applications require an accurate measurement of the intensity of optical radiation. One way to measure the intensity is by absorbing it and measuring the resultant rise in temperature of the absorber. The temperature rise is compared with the temperature rise when the radiation is blocked and the absorber is heated electrically by means of a current I in a resistor R. The power dissipated in R is $I^2 R$ and is accurately measurable. In principle, this is therefore also the radiative power absorbed by the absorber when its temperature rise upon exposure to the radiation equals that caused by the electrical heating. So this 'electric substitution principle' relates optical power to electrical power. A schematic diagram of a radiometer based on this principle is shown in figure 1.2.

A 'room-temperature' radiometer has an accuracy of the order of 0.1%, which is sufficient for many industrial purposes. However, higher accuracy is needed for establishing the SI base unit of luminous intensity, the 'candela',[12] for accurate measurements of black-body radiation and for space applications such as measurements of solar radiation and reflected radiation from the Earth. In 1985, T. J. Quinn and J. E. Martin of the National Physical Laboratory of the UK described the operation of a radiometer at cryogenic temperatures. The main purpose was of this work was the determination of the Stefan–Boltzmann constant, which relates the temperature of a black body to the amount of radiation it

---

[12] Definitions of the base units of the SI, the 'Système International', are listed in chapter 2.

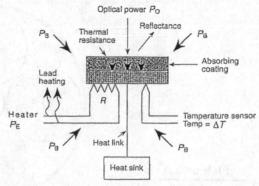

Figure 1.2. A schematic diagram of a radiometer with electric substitution. The absorbing coating is a black paint on the inner surface of a copper cavity (not shown). P_B denotes background radiation. Uncertainties are reduced by a factor of between 10 and 50 when the instrument is operated at cryogenic temperatures (courtesy of *Metrologia*).

emits.[13] In cryogenic radiometers, the working temperature is commonly about 6 K ($-267\,°C$). The absorber is a special black paint over a copper surface and the copper is shaped as a cylindrical cavity (not shown in figure 1.2), with a small aperture through which the radiation enters.

Cryogenic operation confers the following advantages.[14] The leads to the resistor R (a rhodium–iron alloy) can be made of a metal, such as niobium, which becomes a superconductor (with exactly zero resistance) at low temperatures. The electrical heating is then concentrated in R and there is a negligible amount of 'background' heating. Because of the lowered specific heat of copper at cryogenic temperatures, the thermal diffusivity of copper is about a thousand times higher than its room-temperature value. This means that a rise in temperature at any point in the copper very quickly diffuses to the rest of the copper and can be detected by the sensor. In turn this means that the cavity can be large, increasing its efficiency as an absorber of radiation, yet a short response time (of the order of a few minutes) to both optical and electrical heating is obtained. A short response time permits fast optical and electrical cycling. Furthermore, at low temperatures radiative heat loss is reduced and, since the entire apparatus is operated in a vacuum, there is no convective heat loss to the environment. Background radiation (denoted P_B in figure 1.2) is also reduced at low temperatures. The accuracy of the cryogenic radiometer is in the range 0.002% to 0.01%, and this makes it a primary standard for the measurement of intensity of optical radiation.

---

[13] According to the Stefan–Boltzmann law for a black body, the amount, W, of radiation in watts per square metre is related to the absolute temperature, T, by $W = \sigma T^4$. The approximate value of σ is $5.7 \times 10^{-8}\ W\cdot m^{-2}\cdot K^{-4}$.

[14] Instrumentation used in the space applications referred to is calibrated against a ground-based cryogenic radiometer.

1.2 Review

Measurement affects our lives by alerting us to the speed at which we are driving, the amount of luggage we can carry onto an airplane and even the amount of time required to boil an egg. For scientists and engineers accurate measurement is a habitual preoccupation. Metrology is the name given to the science of accurate measurement and of estimation of measurement uncertainties. Metrology, directed by a country's national measurement institute (NMI), underpins the reliability and quality control of industrial products. Careful measurement may mean, for example, that an accepted theory is in need of revision, or that a new design is required in some critical component of an aircraft.

In order to communicate the results of measurements effectively and efficiently, scientists and engineers must agree upon a system of units in which to measure mass, time, length and other physical quantities. A convenient and widely adopted system of units is the SI, and it is this system that we will focus upon in the next chapter.

Further reading

Several measurement topics were discussed in this chapter. The interested reader will find more information in the following references. The references are numbered according to the section in which the topic was mentioned.

(1) D. Kind and H. Lübbig (2003), 'Metrology – the present meaning of a historical term', *Metrologia*, **40**, 255–257.

(1.1) R. Myors, S. Askey and L. Mackay (2004), 'An isotope dilution mass spectrometer method for mercury in fish', *Proceedings of the Metrology Society of Australia*, March, 167–170.

M. J. Turner, B. A. Baker and P. C. Kam (2004), 'Effects of systematic errors in blood pressure measurements on the diagnosis of hypertension', *Blood Pressure Monitoring*, **9** (5), 249–253.

(1.1.1) I. M. Mills, P. J. Mohr, T. J. Quinn, B. N. Taylor and E. R. Williams (2005), 'Redefinition of the kilogram: a decision whose time has come', *Metrologia*, **42**, 71–80.

E. Adelberger, B. Heckel and C. D. Hoyle (2005), 'Testing the gravitational inverse-square law', *Physics World*, April, 41–45.

(1.1.2) On the discovery of argon: Rayleigh described his experiments in Rayleigh, Lord (1895), 'Argon', *Proceedings of the Royal Institution*, **14**, 524. See also http://web.lemoyne.edu/~giunta/rayleigh.html.

(1.1.3) J. C. Farman, B. G. Gardiner and J. D. Shanklin (1985), 'Large losses of total ozone in Antarctica reveal seasonal ClO_x/NO_x interaction', *Nature*, **385**, 207–210.

(1.1.4) T. J. Quinn (2003), 'Practical realization of the definition of the metre, including recommended radiations of other optical frequency standards', *Metrologia*, **40**, 103–133.

P. D. Fisher (1992), 'Improving on police radar', *Institution of Electrical and Electronics Engineers Spectrum*, July, 38–43.

(1.1.5) N. Ashby (2002), 'Relativity and the Global Positioning System', *Physics Today*, May, 41–47.

(1.1.6.1) B. I. Bleaney and B. Bleaney (1976), '*Electricity and Magnetism*, 3rd edn, Oxford, Oxford University Press.

A. M. Thompson (1968), 'An absolute determination of resistance based on a calculable standard of capacitance', *Metrologia*, **4**, 1–7.

A. M. Thompson and D. G. Lampard (1956), 'A new theorem in electrostatics and its application to calculable standards of capacitance', *Nature*, **177**, 888.

Geometry-independence in the calculable capacitor is such that only a single distance needs to be measured. There are cases in electrical metrology where geometry-independence is complete, in the sense that the basic equations determining an electrical standard involve no length measurements at all. Three examples are the Josephson effect as used for voltage standards, the quantum Hall effect as used for resistance standards and the so-called cryogenic current comparator for measuring ratios of direct currents ('dc'). These are described, respectively, in the following three articles:

C. A. Hamilton, C. Burroughs and K. Chieh (1990), 'Operation of NIST Josephson array voltage standards', *Journal of Research of the National Institute of Science and Technology*, **95**, 219–235.

B. Jeckelmann and B. Jeanneret (2001), 'The quantum Hall effect as an electrical resistance standard', *Reports on Progress in Physics*, **64**, 1389–1441.

I. K. Harvey (1976), 'Cryogenic a.c. Josephson effect emf standard using a superconducting current comparator', *Metrologia*, **12**, 47–54.

(1.1.6.2) T. J. Quinn and J. E. Martin (1985), 'A radiometric determination of the Stefan–Boltzmann constant and thermodynamic temperatures between $-40\,°\mathrm{C}$ and $+100\,°\mathrm{C}$', *Philosophical Transactions of the Royal Society*, **316**, 85–181.

N. P. Fox (1995–1996), 'Radiometry with cryogenic radiometers and semiconductor photodiodes', *Metrologia*, **32**, 535–543.

The analogous electric substitution principle for comparing constant (direct-current or dc) voltages and alternating ('ac') voltages is described in

B. D. Inglis (1992), 'Standards for ac–dc transfer', *Metrologia*, **29**, 191–199.

2

Measurement fundamentals

Lord Kelvin, a renowned scientist born in Ireland in the nineteenth century, recognised the importance of measurement and spoke about it in passionate terms:

When you can measure what you are speaking about and express it in a number, you know something about it; but when you cannot measure it, when you cannot express it in numbers, your knowledge is of a meagre and unsatisfactory kind; it may be the beginning of knowledge, but you have scarcely in your thoughts advanced to the state of science ...
(Lecture given to the Institution of Civil Engineers, 3 May 1883)

Measurement is essential to science. Without measurement, scientific models and theories cannot be rigorously tested or challenged. Ambitious scientific and technical endeavours such as the exploration of the surface of Mars, medical diagnosis using magnetic-resonance imaging (MRI) and the evaluation of renewable energy sources would not be possible. Measurement is no less critical in areas such as international trade, with the global economy becoming ever more pervasive.

In this chapter we consider matters key to measurement and the communication of the results of measurement. These include the system of units, scientific notation and significant figures.

2.1 The system of units of measurement

To measure the length of a particular object and have the result of that measurement recognised and understood by other people, there must be mutual agreement on a basic unit of length. Over past centuries many units have been adopted as the basis of length measurement in different parts of the world. Some of those units, like the metre, the mile and the fathom, are still in use today. Other units of length such as the barleycorn and the ell are extinct. In order to simplify the measurement of length and to avoid the need to remember factors needed to convert, say, a distance expressed in miles to a distance expressed in metres, it seems sensible (for the science and engineering communities at least) to agree to use a single unit for length measurement. This permits clear communication of results of measurements of length. Equally, simplification and standardisation in measurement bring economic gains when products and services that rely on length measurement are traded between states and countries.

Length is only one of several characteristics of an object that we might wish to quantify – that is, attach a numerical value to. Others include the object's mass, temperature and electrical resistance. To quantify these as well as other characteristics, we need a system of units that allows for the measurement of any physical quantity. Specifically, we require a system of units that is

• comprehensive,
• internationally accepted and adopted,[1]
• coherent and convenient to use and
• regularly reviewed and revised.

The most common system of units used worldwide in science and technology is the Système International – commonly referred to as the SI.

2.1.1 The SI

The origin of the SI can be traced back to the late eighteenth century in France when the metre was specified as the distance between two marks on a platinum bar.[2] The kilogram was defined as the mass of water filling a cube one-tenth of a metre on a side and, like the metre, was constructed as a platinum artefact. Together with the unit of time, the 'second', defined as 1/86 400 of the mean solar day (in 1960 it was redefined in terms of a tropical year), these three units were the earliest of the system of seven base units now known as the SI. With advances in science, the definitions of the metre and of the second have changed (see table 2.1).

The Metre Convention, signed in Paris in 1875 by representatives of 17 nations, established three international bodies in metrology. These are (with their French acronyms) the General Conference on Weights and Measures (CGPM), the International Committee for Weights and Measures (CIPM) and the BIPM. The CGPM

---

[1] A few centuries ago, units of measurement were variable and inconsistent to a degree that would now be considered intolerable. The 'ell', a unit of length roughly that of the adult human outstretched arm, was about 20% shorter in Scotland than in England (Klein 1989). As late as the 1850s the 'Pfund', a unit of mass, was almost 1.6% larger in Berne than in Zurich (Barnard 1872).

[2] This was not an arbitrary distance. In 1791 the metre was defined as one ten-millionth of the meridian distance between the North Pole and the equator, passing through the cities of Dunkirk and Barcelona, whose latitudes were accurately known. The approximately 1100-km distance between the cities was measured by J. Delambre and P. Méchain in a monumental project that lasted from 1792 to 1799, and carried the additional hazard of coinciding with the French Revolution. Nevertheless, the value obtained for the pole–equator distance was only about 0.02% different from the currently agreed value (Alder 2002).

Table 2.1. *SI base units, symbols and definitions*

Quantity	Unit	Symbol	Definition
Mass	kilogram	kg	The kilogram is equal to the mass of the international prototype of the kilogram. (The prototype kilogram is made from an alloy of platinum and iridium and is kept under carefully controlled environmental conditions near Paris.)
Time	second	s	The second is the duration of 9 192 631 770 periods of the radiation corresponding to the transition between the two hyperfine levels of the ground state of the cesium-133 atom.
Length	metre	m	The metre is the length of the path travelled by light in a vacuum during a time-interval of 1/299 792 458 of a second.
Thermodynamic temperature	kelvin	K	The kelvin is the fraction 1/273.16 of the thermodynamic temperature of the triple point of water.
Electric current	ampere	A	The ampere is that current which, if maintained in two straight parallel conductors of infinite length, of negligible cross-section and placed one metre apart in a vacuum, would produce between these conductors a force of 2×10^{-7} newton per metre of length.
Luminous intensity	candela	cd	The candela is the luminous intensity, in a given direction, of a source that emits monochromatic radiation of frequency 540×10^{12} hertz and that has a radiant intensity in that direction of 1/683 watt per steradian.
Amount of substance	mole	mol	The mole is the amount of substance of a system which contains as many elementary entities as there are atoms in 0.012 kilogram of carbon 12.

is the ultimate custodian of the SI. Research in metrology performed at the BIPM and other NMIs around the world is reported to the CIPM through a system of consultative committees of experts drawn from NMIs. Any improvements or changes to the SI, or changes to the as-maintained units that might follow from fundamental research reported to the CIPM, are decisions made by the CGPM. Thus, for example, the as-maintained unit of electromotive force or electric potential, the volt, was changed on 1 January 1990 (by about eight parts per million) as a consequence of painstaking 'absolute' measurements of the volt (that is, measurements of the force of electrostatic attraction generated by a voltage).[3] At present (2005) 51 nations are signatories to the Metre Convention.

2.1.2 Base and derived units

The SI consists of seven base units. These units, their symbols and their definitions are shown in table 2.1. It should be noted that the speed of light in a vacuum, c, has a defined quantity, $299\,792\,458\ \mathrm{m} \cdot \mathrm{s}^{-1}$, with zero uncertainty.

Other units, created by combining SI base units, are referred to as *derived* units. As an example, the average speed, v, of a body is related to the distance, d, travelled by a body in a time, t, through the equation

$$v = \frac{d}{t}. \tag{2.1}$$

Replacing each quantity on the right-hand side of equation (2.1) by its unit gives

$$\text{unit of speed} = \frac{\mathrm{m}}{\mathrm{s}}, \text{ which may also be written as either m/s or } \mathrm{m} \cdot \mathrm{s}^{-1}.$$

The unit of speed does not have a special name, but there are derived units, such as those of force and energy, that do. Table 2.2 contains examples of SI derived units that have special names.

The units of other quantities, such as latent heat, are usually expressed as a combination of derived units, with special names, and base units. Examples of such combinations are shown in table 2.3. While it is often convenient to use derived units with special names when indicating the units of a quantity, all units may be expressed in terms of base units.

Example 1
Show that the derived unit N/m (which is the unit of, for example, surface tension) can be expressed in base units as $\mathrm{kg} \cdot \mathrm{s}^{-2}$.

---

[3] See section 4.1.3 and footnote 6 in chapter 4.

Table 2.2. *Examples of derived units with special names*

Quantity	Derived unit	Symbol	Unit of quantity expressed in base units
Frequency	hertz	Hz	s^{-1}
Force	newton	N	$kg \cdot m \cdot s^{-2}$
Pressure	pascal	Pa	$kg \cdot m^{-1} \cdot s^{-2}$
Energy, work	joule	J	$kg \cdot m^2 \cdot s^{-2}$
Power	watt	W	$kg \cdot m^2 \cdot s^{-3}$
Potential difference, electromotive force (emf)	volt	V	$kg \cdot m^2 \cdot s^{-3} \cdot A^{-1}$
Electrical charge	coulomb	C	$s \cdot A$
Electrical capacitance	farad	F	$kg^{-1} \cdot m^{-2} \cdot s^4 \cdot A^2$
Electrical resistance	ohm	Ω	$kg \cdot m^2 \cdot s^{-3} \cdot A^{-2}$
Electrical conductance	siemens	S	$kg^{-1} \cdot m^{-2} \cdot s^3 \cdot A^2$
Magnetic flux density	tesla	T	$kg \cdot s^{-2} \cdot A^{-1}$
Magnetic flux	weber	Wb	$kg \cdot m^2 \cdot s^{-2} \cdot A^{-1}$
Inductance	henry	H	$kg \cdot m^2 \cdot s^{-2} \cdot A^{-2}$
Absorbed dose	gray	Gy	$m^2 \cdot s^{-2}$
Reaction rate	katal	kat	$mol \cdot s^{-1}$

Table 2.3. *Examples of other derived units incorporating units with special names*

Quantity	Derived unit	Symbol	Unit of quantity expressed in base units
Specific heat capacity	joule per (kilogram kelvin)	J/(kg · K)	$m^2 \cdot s^{-2} \cdot K^{-1}$
Thermal conductivity	watt per (metre kelvin)	W/(m · K)	$kg \cdot m \cdot s^{-3} \cdot K^{-1}$
Latent heat	joule per kilogram	J/kg	$m^2 \cdot s^{-2}$
Electric field strength	volt per metre or newton per coulomb	V/m or N/C	$kg \cdot m \cdot s^{-3} \cdot A^{-1}$
Molar entropy	joule per (mole kelvin)	J/(mol · K)	$kg \cdot m^2 \cdot s^{-2} \cdot mol^{-1} \cdot K^{-1}$
Radiance	watt per (square metre steradian)	W/(m² · sr)	$kg \cdot s^{-3}$
Electrical resistivity	ohm metre	$\Omega \cdot m$	$km \cdot m^3 \cdot s^{-3} \cdot A^{-2}$

Answer

From table 2.2, $N = kg \cdot m \cdot s^{-2}$, so

$$\frac{N}{m} = \frac{kg \cdot m \cdot s^{-2}}{m} = kg \cdot s^{-2}.$$

Table 2.4. *Prefixes used with the SI*

Factor	Prefix	Symbol	Factor	Prefix	Symbol
10^{-24}	yocto	y	10^{1}	deka	da
10^{-21}	zepto	z	$\mathbf{10^{2}}$	**hecto**	**h**
$\mathbf{10^{-18}}$	**atto**	**a**	$\mathbf{10^{3}}$	**kilo**	**k**
$\mathbf{10^{-15}}$	**femto**	**f**	10^{6}	mega	M
$\mathbf{10^{-12}}$	**pico**	**p**	10^{9}	giga	G
$\mathbf{10^{-9}}$	**nano**	**n**	10^{12}	tera	T
$\mathbf{10^{-6}}$	**micro**	**μ**	10^{15}	peta	P
$\mathbf{10^{-3}}$	**milli**	**m**	10^{18}	exa	E
10^{-2}	centi	c	10^{21}	zetta	Z
10^{-1}	deci	d	10^{24}	yotta	Y

Exercise A

With the aid of table 2.2, write the following units in terms of base units only:

(a) F/m, (b) W/m^2, (c) J/m^3, (d) J/K, (e) $\Omega \cdot$ m, (f) Ω/m^2, (g) C/kg, (h) Wb/A, (i) C^2/(N $\cdot$ m^2), (j) N $\cdot$ m, (k) N/A^2, (l) W/(m$^2 \cdot$ K^4)

The SI units have the great advantage of being *coherent*. This means that any theoretically derived equation relating physical or chemical quantities is automatically satisfied numerically if all the quantities are simultaneously expressed in the SI units of table 2.1 or the derived units such as those in tables 2.2 and 2.3. For example, the equation for kinetic energy E of a mass m, moving at velocity v, is given by $E = \frac{1}{2}mv^2$. If m is given as a value in kilograms and v in metres per second, then E is automatically the correct value for the kinetic energy in joules.

2.1.3 Prefixes

A quantity, such as a time interval, may span many orders of magnitude. At one extreme we might need to consider the time interval taken for an electromagnetic wave to travel the distance equal to the diameter of the nucleus of hydrogen. At the other extreme, we might require an estimate of the age of the Universe. These and other time intervals between these extremes may be expressed by multiplying the unit of time, i.e. the second, by an appropriate power of ten. This is indicated succinctly by attaching a prefix to the unit. For example, one thousandth of a second is expressed as 1 ms, where m stands for milli, equivalent to 1/1000. The SI uses prefixes that represent multiplying factors covering the range 10^{-24} to 10^{24}. Table 2.4 includes the prefixes currently used in the SI. Some prefixes are better known and more frequently used than others. The more frequently used prefixes are indicated in bold in table 2.4.

In practice, some prefixes are seldom used in conjunction with particular units. For example, while the centimetre (cm) is regularly used as a unit of length, the centinewton (cN) is rarely encountered. Similarly, while electrical resistance is often expressed in megohms (MΩ), it is rare to find time expressed in megaseconds (Ms). It should be emphasised here that 'kilogram', although it has the prefix 'kilo', is an SI base unit and is the only SI base unit with a prefix. As a cautionary note, the coherence of the SI units, referred to above, does not automatically extend to the case where the units have prefixes (with the exception of the kilogram). Thus, for $E = \frac{1}{2}mv^2$, if m were measured in grams or v in kilometres per second, E would not be automatically obtained in joules.

Example 2
Express the following values with the aid of SI prefixes:

(a) 3.4×10^{-3} A,   (b) 6.4×10^{-5} m^2/s,   (c) 7.5×10^8 Ω,   (d) 1.8×10^{10} Pa,
(e) 3.5×10^5 $\Omega \cdot$ m

Answer
Although there are no restrictions on the use of prefixes, it is usual (and rational) to choose a prefix similar in magnitude to the value being considered:

(a) 3.4 mA, (b) 64 mm^2/s, (c) 0.75 GΩ, (d) 18 GPa, (e) 0.35 M$\Omega \cdot$ m

Exercise B
Express the following values using the most appropriate SI prefixes:

(a) 7.7×10^{-9} C, (b) 0.52×10^{-10} J, (c) 7834 V, (d) 1.3×10^7 m/s, (e) 3.5×10^{-4} Pa $\cdot$ s

2.2 Scientific and engineering notations

Many values are conveniently expressed using a number, a prefix and an SI unit. For example, the mass of a small body may be expressed as 65 mg. The same value can be expressed using powers-of-ten notation. In fact, in situations in which prefixes are unfamiliar, it is perhaps preferable to adopt 'powers-of-ten' notation. As an example, few would immediately recognise 0.16 aC as the magnitude of the charge carried by an electron. By contrast, expressing the same value as 1.6×10^{-19} C is likely to bring a nod of recognition from many working in science.

A value of length, such as $l = 13\,780$ m, is expressed in scientific notation using the following steps. Separate the first non-zero digit from the second by a decimal point, such that 13 780 becomes 1.3780. Now multiply this number by ten raised to the appropriate power in order to return the number back to its original magnitude.

In this example this is 10^4. The value is now written (not forgetting to include the unit) as

$$l = 1.3780 \times 10^4 \, \text{m}$$

Exercise C

Write the following values using scientific notation:

(a) 0.0675 N, (b) 3000 kg, (c) 160 zC, (d) 755 mV, (e) 0.0035 kat, (f) 982.1 MW

Engineering notation differs from scientific notation in that the powers of ten that follow the multiplication sign are limited to $3n$ where $n = 0, \pm 1, \pm 2$ etc. For example, the value $l = 13\,780$ m would be written in engineering notation as

$$l = 13.780 \times 10^3 \text{m}.$$

Exercise D

Express the values in exercise C in engineering notation.

2.3 Rounding and significant figures

When a number has too many significant figures for a particular purpose, the number of significant figures can be reduced by a simple procedure known as 'rounding'. For example, if a distance has been measured as 1.1451 m, that is to five significant figures, but three significant figures are considered sufficient, that value can be rounded to 1.15 m. We note that 1.14 m would be incorrect rounding, since 1.15 is closer to 1.1451 than 1.14 is. If 1.1451 m is to be rounded to two significant figures, this gives 1.1 m.

The following question arises: what if the original figure ends in '5'? The recommended rounding advice is as follows: choose the even round value. Thus we have, for example,[4]

3.05 is rounded to 3.0
3.15 is rounded to 3.2
3.25 is rounded to 3.2
3.35 is rounded to 3.4
3.45 is rounded to 3.4

and so on. This has the advantage that dividing both unrounded and rounded values by 2 still gives the correct relationship; for example, dividing 3.05 by 2 gives 1.525, which rounds to 1.5.

---

[4] For more detail and discussion, see Australian Standard AS2706-2003, 'Numerical values – rounding and interpretation of limiting values' (Sydney, Standards Australia, 2nd edn, 2003).

When measurements are made, how many figures should be reported? Modern instruments are capable of displaying values to many figures. As an example, a $5\frac{1}{2}$-digit digital multimeter (DMM) on its 2-V d.c. range can indicate any value between -1.99999 V and $+1.99999$ V. While it is often prudent to record all the figures supplied by an instrument, in many cases the particular quantity being measured may vary to such an extent as to make some of the figures almost meaningless.

Let us consider the situation in which a $5\frac{1}{2}$-digit voltmeter is used to measure the output of an optical transducer. At one instant the voltmeter displays a value of 1.675 43 V and a half-second later the display indicates 1.652 13 V. In the absence of a statistical analysis on many repeat values, such as that discussed in chapter 5, we might decide to round the voltages to 1.68 V and 1.65 V in recognition of the fact that the last three figures in the display were unreliable and of little use. Then 1.68 V and 1.65 V are values expressed to *three significant figures*. The fact that a statistical analysis was not carried out in order to assess the uncertainty in the value[5] means that to a certain extent the rounding to three significant figures was arbitrary and represents the experimenter exercising 'common sense'. When experimental values with varying numbers of significant figures are brought together, there are several simple rules that allow us to quote answers to a defensible number of significant figures. It is emphasised that, although these rules are helpful, they are not a substitute for the detailed calculation of uncertainty to be described in the following chapters (specifically, chapters 7 onwards). Such a calculation indicates how many figures should be used when quoting a value.

Rule 1

In the absence of any explicit statement about the uncertainty of a quoted value, the approximate uncertainty in a value can be estimated as half the possible range of the values with an extra decimal place that are all consistent, after rounding, with the quoted value.

Suppose that a distance, d, is quoted as 25.1 m. This implies a possible distance anywhere in the approximate interval 25.05 m to 25.15 m. This interval comprises an infinite number of values, all of which (except for the first and last, following the rounding advice above) would be 'rounded' to the quoted 25.1 m and so are consistent with 25.1 m. The interval containing these values is $(25.15 - 25.05)$ m $= 0.10$ m. Half this interval is 0.05 m. So we infer from the quoted value of distance $d = 25.1$ m that the uncertainty is 0.05 m. The *proportional* uncertainty is then $0.05/25.1$ or about five parts in 2500 or 0.2%.

---

[5] Such an analysis is to be preferred, since it leads to a clearer decision as to how many figures to retain.

Rules 2 and 3 say, essentially, that the proportional uncertainty in the result of a calculation is dominated by the *least* accurate component in the calculation.

Rule 2

When values are multiplied or divided, quote the answer to the number of significant digits that implies a proportional uncertainty closest to the greater of the component proportional uncertainties.

For example, suppose that we require the value of a speed v when the distance $d = 25.1$ m and the time taken is $t = 3.4$ s. The distance d is quoted to one part in 500 (0.2%) as discussed above, and the time t to roughly five parts in 300 or one part in 60 (1.7%). We write, provisionally,

$$v = \frac{d}{t} = \frac{25.1 \text{ m}}{3.4 \text{ s}} = 7.382\,353 \text{ m/s}.$$

The greater of the component proportional uncertainties is evidently one part in 60 or 1.7%. It is useful to make up a table of possible quoted values and the resulting implied proportional uncertainties:

Quoted value	Implied proportional uncertainty
7.38 m/s	5 parts in 7000 or 0.07%
7.4 m/s	5 parts in 700 or 0.7%
7 m/s	5 parts in 70 or 7%

Of these possibilities, we should choose 0.7% as closest to the required 1.7%, and so we quote the speed as 7.4 m/s.

Rule 3

When numbers are added or subtracted, quote the answer to the number of significant digits that implies a proportional uncertainty closest to the greater of the component proportional uncertainties.

As an example, if the mass of a copper container filled with water is given as $m_{Cu+H_2O} = 1.5778$ kg and the mass of the empty copper container is given as $m_{Cu} = 0.562$ kg, then the mass of the water in the container, m_{H_2O}, is given by

$$m_{H_2O} = m_{Cu+H_2O} - m_{Cu} = 1.5778 \text{ kg} - 0.562 \text{ kg} = 1.0158 \text{ kg}.$$

The mass quoted as 1.5778 kg implies a proportional uncertainty of about five parts in 160 000 or 0.003%. The mass quoted as 0.562 kg implies a proportional uncertainty of roughly five parts in 6000 or 0.08% and is the higher of the two proportional uncertainties. We therefore require the mass of water, 1.0158 kg, to be quoted with an implied proportional uncertainty as close as possible to 0.08%. Here is a table of possible quoted values and the resulting implied proportional uncertainty:

Quoted value	Implied proportional uncertainty
1.0158 kg	5 parts in 100 000 or 0.005%
1.016 kg	5 parts in 10 000 or 0.05%
1.02 kg	5 parts in 1000 or 0.5%
1.0 kg	5 parts in 100 or 5%.

It is clear that 0.05% proportional uncertainty is the closest to the required 0.08%, and so we quote the mass of water as 1.016 kg.

Rule 4

When a quantity with proportional uncertainty p is raised to the power n, the resultant proportional uncertainty is $|np|$ and the quoted number of significant figures should reflect this.

For example, the formula for the volume V of a sphere of diameter D is $V = \frac{1}{6}\pi D^3$. Suppose that D is given as 50.1 mm. This implies a proportional uncertainty of five parts in 5000 or 0.1%. The calculated value of V is 65 843 mm^3 and should be quoted with significant digits implying 0.3% proportional uncertainty. We have the following table:

Quoted value	Implied proportional uncertainty
6.584×10^4 mm^3	5 parts in 66 000 or 0.008%
6.58×10^4 mm^3	5 parts in 6600 or 0.08%
6.6×10^4 mm^3	5 parts in 700 or 0.7%
7×10^4 mm^3	5 parts in 70 or 7%

The proportional uncertainty closest to 0.3% is 0.7%, implying a quoted value for the volume of 6.6×10^4 mm^3. We note that writing this as 66 000 mm^3 implies too low an uncertainty, in view of the zeros. Finally, it is worth noting that, if n is between -1 and 1, implying a fractional power, the proportional uncertainty in the result will be less than that in the original data. This would arise, for example, if we calculated the diameter of a sphere when given its volume.

Exercise E

Apply rules 1, 2 and 3 to give the outcomes of the following calculations to an appropriate number of significant figures:

(a) 2.343 m/s $\times$ 1.52 s, (b) $\dfrac{2.3 \times 10^{-16}\,\text{J}}{1.602 \times 10^{-19}\,\text{C}}$, (c) 1.5751 g + 10.27 g,

(d) $\dfrac{1.22 \times 10^{-20}\,\text{J}}{1.38 \times 10^{-23}\,\text{J/K} \times 273.15\,\text{K}}$

Where there are several steps in a calculation, it is prudent not to round intermediate results, since premature rounding introduces unnecessary error that will propagate through to the final answer. If in doubt, use all the figures available at each step in a calculation. Once the final answer has been obtained, review the calculations to determine which values are given to the least number of significant figures, then round accordingly.

2.4 Another way of expressing proportional uncertainty

Proportional uncertainties are often expressed as percentages, as in the examples in Section 2.3. When very accurate measurements are made, the uncertainties may be expressed in parts per million or even parts per billion. An error of, say, 3 micrometres when 1 metre is being measured is expressible as three parts per million. It is often advantageous to retain information on the physical quantity being measured, so we may express three parts per million in the measurement of a metre as 3 μm/m. Similarly, if a constant voltage of 2 V is being measured with an uncertainty of 5 μV or 2.5 parts per million, this may be expressed as an uncertainty of 2.5 μV/V.

2.5 Review

In order to communicate the result of a measurement we must assign a number and a unit to values emerging from an experiment. In this chapter we have considered units of measurement with particular emphasis on the SI. To express very large and very small values, it is convenient to use prefixes, such as giga and micro, or to adopt scientific or engineering notation. We introduced rules for presenting numbers to a plausible number of significant figures, although these rules can be set aside once a complete error analysis such as that indicated in chapter 10 has been undertaken.

Analysis of errors requires that we use some terms that have an 'everyday' meaning, such as accuracy and precision. With a view to clarifying the situation, the next chapter focusses on frequently used terms in measurement.

3

Terms used in measurement

For the newcomer, unfamiliarity with the specialist vocabulary of scientific disciplines like physics and chemistry can act as an obstacle to learning those disciplines. What can be even more challenging is that science employs many words such as *force* and *energy* that are used in various ways in everyday language. The science of measurement, in particular, has many terms, such as error, uncertainty and accuracy, that also occur in day-to-day use in contexts far removed from measurement. In this chapter we consider terms used in measurement, including those with an everyday or popular meaning such as *error*, and we clarify their meaning when used in the context of measurement.[1]

3.1 Measurement and related terms

3.1.1 Measurement

Measurement is a process by which a value of a particular quantity such as the temperature of a water bath or the pH of a solution is obtained. In the case of length measurement, this might involve measuring the atomic-scale topography of a surface using an instrument such as an atomic-force microscope (AFM), or measuring the length of a pendulum using a metre rule. Values obtained through measurement form the foundation upon which we are able to

- test both new and established scientific theories;
- decide whether a component, such as a resistor, is within specification;

---

[1] Because it is important to define terms clearly in order to avoid ambiguity, the International Standardisation Organisation (ISO), representing the international measurement community, has published a document called *International Vocabulary of Basic and General Terms in Metrology*, known as the VIM for short (www.iso.org). Here we focus on terms in the VIM that are most commonly used.

- compare values obtained by workers around the world of a particular quantity, such as the thickness of the ozone layer of the atmosphere;
- quantify the amount of a particular chemical species, such as the amount of steroid in a sample of urine taken from an athlete; and
- establish the proficiency of laboratories involved with the testing and calibration of equipment.

3.1.2 Measurands

A particular quantity determined through measurement is called a *measurand*. Often a value of a measurand can be established directly with an instrument. For example, the period, T, of a pendulum, can be measured using a stopwatch. Determining the period of the pendulum may be an end in itself, or it may be used in the determination of another measurand. For example, by combining the period with the length of the pendulum, l, the acceleration caused by gravity, g, when a body falls freely can be determined by means of the equation

$$g = \frac{4\pi^2 l}{T^2}.$$

Though g is not obtained directly, in the sense that an instrument does not indicate its value, it too is a measurand.

Some care is required when describing a measurand. For example, the purpose of a measurement may be to determine the density of a metal such as platinum. The density of any metal sample depends on the purity of the metal as well as on its temperature. Unless the description of the measurand includes specification of purity as well as temperature, reported values of densities of samples of platinum are likely to vary significantly from one observer to the next.

3.1.3 Units and standards

A properly defined unit allows quantities of the same kind to be compared. While there are several systems of units in existence, the most widely used system in science and engineering is the SI.

For example, if we have a rule that is one metre long, then any length can be compared with that rule and can be expressed as a multiple of one metre. For some comparisons it is important that a 'standard' rule be used that is as accurate and stable as possible. One of the duties of NMIs is to house, safeguard and maintain the highest-quality standards, referred to as *primary standards*.

Figure 3.1. The Australian 1-kilogram standard, copy 44 of the BIPM 1 kilogram. The material is a platinum–iridium alloy and the cylinder has diameter equal to its height, each about 39 mm. The material is about 2.7 times denser than steel, implying a small volume for 1 kg and therefore a small buoyancy correction. The alloy has high resistance against corrosion and good electrical and thermal conductivities (Davis 2003).

The kilogram is the only SI base unit[2] that is realised by means of an artefact.[3] This primary standard is made of a platinum–iridium alloy and is kept under very tightly controlled conditions, together with several copies, at the BIPM in Paris. Other copies are held at NMIs around the world; the Australian kilogram standard is Copy No. 44 of the BIPM kilogram (see figure 3.1) and is kept at the National Measurement Institute of Australia in Sydney.

All other SI base units are defined through constants of nature, so that any well-equipped NMI in the world can realise the base unit. For example, the realisation of the metre is accomplished using a laser whose light has a precise frequency, f, that

---

[2] The kilogram may eventually be defined in terms of natural constants. One method would be to define it as the mass of a specified number of atoms of a particular isotope of an element. There is considerable progress along this route, which involves the fabrication of very accurate spheres of pure silicon of accurately known diameter. The lattice spacing of silicon atoms in this structure is well known (from X-ray-crystallographic measurements) and, since the volume is also well known (given the diameter and the accurate sphericity), the number of atoms can in principle be counted. Since the relative proportions of the three stable isotopes of silicon can be measured accurately, this number of atoms then determines the total mass. Another method, also well advanced, would assign a defined value to the Planck constant h. This would have the effect of defining the kilogram. Both methods and their metrological consequences are discussed in Mills *et al.* (2005).

[3] 'Artefact', in this context, has a different meaning from 'artefact' with a negative connotation that describes an 'artificial' or anomalous value, affected by some extraneous influence, in a series of measured values.

Figure 3.2. A 1-Ω standard resistor and its protective case (courtesy B. J. Pritchard, National Measurement Institute of Australia).

is determined by atomic processes, is largely independent of human intervention, and can be measured in terms of the second as defined in table 2.1 of chapter 2. Using the defined and uncertainty-free value for the speed of light c, the metre can then be defined as a multiple of the wavelength, λ, inferred from f through the relation $\lambda = c/f$.

3.1.4 Calibration

In order that an instrument or artefact should accurately indicate the value of a quantity, the instrument or artefact requires *calibration*. This procedure is essential for establishing the traceability of the instrument or artefact to a primary standard (see section 3.1.5).

There is no hard-and-fast distinction between 'instrument' and 'artefact', but in general an instrument *measures* a quantity, whereas an artefact *provides* a quantity. For example, a digital multimeter (DMM) is an instrument that measures voltage, resistance or current and displays it as a number. An instrument may also measure a quantity by means of the position of a pointer on a dial.[4] By contrast, standard weights and gauge blocks are artefacts, also known as artefact standards or standard artefacts. Figure 3.2 shows a very stable standard of electrical resistance: a

---

[4] The position of a pointer on a dial may be regarded as an 'analogue' display, in contrast to a digital display, which is usually more accurate. When quickness of reading is more important than accuracy, analogue displays are preferred. This is why vehicle speedometers are usually analogue displays.

1-Ω standard artefact, with low temperature coefficient of resistance, which was designed and manufactured at the National Measurement Institute of Australia (Pritchard 1997).

During calibration, a value measured by an instrument or provided by an artefact is compared with that obtained from a standard instrument or artefact. If there is a discrepancy between the value as indicated by the instrument or artefact and the corresponding standard, then the difference between the two is quoted as a correction to the instrument or artefact. This process is referred to as *calibration*, and the correction always has a stated associated uncertainty. Over time it is possible for the values indicated by an instrument or provided by an artefact to 'drift'. This makes recalibration necessary. Manufacturers often advise that calibration be carried out at regular intervals (say every 12 months).

3.1.5 Traceability

The result of a measurement is said to be *traceable* if, through an unbroken chain of comparisons often involving working and secondary standards, the result can be compared with a primary standard. Any instrument or artefact used as part of the measurement process must recently have been calibrated by reference to a standard that is traceable to a primary standard. A requirement of traceability is that the chain of comparisons be documented. The consequences of lack of traceability, in some instances, can be severe. For example, if a component manufacturer cannot satisfy a regulatory authority that results of measurements on its components can be traced back to a primary standard, then that manufacturer may be prohibited from selling its products in its own country or elsewhere.

3.1.6 Value

The process of measurement yields a *value* of a particular quantity. As examples,

- the value of the period of a pendulum, $T = 2.37$ s;
- the value of the length of a pendulum, $l = 1.35$ m; and
- the value of the mass of a steel ball, $m = 67.44$ g.

A value may be regarded as the product of a number and the unit in which the particular quantity is measured.

3.1.7 The true value and best estimate of the true value

Through careful measurement we seek to estimate the true value of a quantity. An experiment might be devised to find the amount of charge carried by an electron.

Evidence indicates that every electron carries the same amount of charge – but what is its value? In this case (and many others) we assume that there is an 'actual' or *true* value of a quantity, such as the value of the charge of an electron. It is the true value that we would like to establish through measurement. We are forced to admit that instruments, no matter how sophisticated or expensive, are imperfect. Sometimes the quantities that we measure may vary slightly over the period of the measurement. Outside influences such as fluctuations in temperature may affect the measuring instruments and the measurand. These factors conspire to prevent us from finding the true value of a quantity that we seek through measurement.

Though we are unable to find the true value of a quantity through measurement, we are able to obtain an *estimate* of the true value. When only random sources act to influence the values obtained, the *best estimate* of the true value is usually taken to be the mean, $\bar{x}$, of n values, where

$$\bar{x} = \frac{\sum_{i=1}^{n} x_i}{n} \tag{3.1}$$

and x_i is the ith value obtained through measurement.

3.1.8 Error (in measurement), including random and systematic error

The term *error* is arguably a greater source of confusion than any other term encountered when discussing measurement. In everyday language it is commonly used to refer to a mistake or a blunder. In the context of measurement, error is defined as the difference between the measured value and the true value:

$$\text{error} = \text{measured value} - \text{true value.} \tag{3.2}$$

The true value of the quantity being measured cannot be known, so it follows that the error as defined by equation (3.2) is also unknowable. It is recognised that sources of error fall into two categories, depending upon how they affect measurement.

In some cases the influences that affect the measurement process, or the quantity being measured, cause values to be randomly distributed above and below the true value. The errors caused by such influences are termed *random errors*. Consider the time measured 'by hand' for an object to fall freely through a fixed distance. Sources of errors include the inconsistent synchronisation of starting the stopwatch as the ball is released and stopping the stopwatch as the object reaches the ground. If the object is released many times and values for the time of fall are accumulated, it is likely that a pattern in the values will emerge. The point here is that some values will lie above the true value and others below. From equation (3.2) this indicates that an error could have a positive or negative sign, with neither sign being favoured.

Pursuing the example of the timing of a falling object a little further, it is possible that the stopwatch has developed a minor fault such that it consistently indicates that the time of fall is slightly greater than the true value. In this case the error, as calculated using equation (3.2), would be consistently positive. This type of error is referred to as *systematic*, since it causes all measured values to be consistently under- or over-estimated.

To a greater or lesser degree, random and systematic sources of error affect all measurements, but whether it is random or systematic errors that dominate in any given situation is sometimes difficult to establish.

3.1.9 Accuracy and precision

A value obtained through measurement may or may not be close to the true value. In situations where we believe that the measured value *is* close to the true value, we say that the measured value is *accurate*.

Since we cannot know whether a value is close to the true value, it is impossible to quantify accuracy. Nevertheless, it is reasonable to assess the methods used to measure a particular quantity and judge one method capable of better accuracy than another. As an example, the determination of the time interval between two events using an automatically triggered electronic counter is likely to be more accurate than measuring the same time interval with a hand-held stopwatch.

When values obtained by repeat measurements of a particular quantity exhibit little variability, we say that those values are *precise*. Precision, like accuracy, is a qualitative term. It is used to convey a sense of the scatter of values when repeat measurements of a particular quantity are made. Values that exhibit little scatter may, owing to the influence of systematic error, be far from the true value. Care must be exercised when measurements are precise since, if a systematic error has not been accounted for, all the values could be misleading. We note that high accuracy implies high precision, but the reverse does not hold: high precision does not imply high accuracy if there exists a significant systematic error.

3.1.10 Uncertainty

Errors are key and unavoidable ingredients of the measurement process. Their net effect is to create an *uncertainty* in the value of a measurand. As with the word 'error', the word 'uncertainty' is used widely in everyday language, such as 'There is some uncertainty as to whether it will rain today.' When used in the context of measurement, uncertainty has a number and (most often) a unit associated with it. More specifically, measurement uncertainty has the same unit as the measurand. The manner by which an uncertainty is calculated depends on the circumstances, but it is usual to apply established statistical methods in order to calculate uncertainty.

A convenient way of expressing the best estimate of the particular quantity as well as the uncertainty in measurement is as follows. Suppose that the best estimate of the period of a pendulum is 2.25 s and the uncertainty is 0.05 s. We write

$$\text{Period of the pendulum} = (2.25 \pm 0.05) \text{ s}.$$

It is inferred from the way the period is written that the true value of the period has a 'good chance' or high probability of lying in the interval between 2.20 s and 2.30 s. In chapter 10 we will consider in some detail the use of the $\pm$ sign and what is meant by 'good chance'.

3.1.11 Repeatability

In many circumstances, measurements are made under (as far as is possible) identical conditions. When this happens, it is possible for the values obtained to exhibit little variation or scatter. In such cases we speak of measurements being *repeatable*.

3.1.12 Reproducibility

Experimenters at different locations around the world need to compare their measurements with other experimenters. If the measurand is well defined there is an expectation that, wherever a measurement is made and whatever techniques are used, the same value should be obtained for the measurand irrespective of who makes the measurement and which instrument is used. If there *is* consistency between values obtained by different experimenters, we say that the value is *reproducible*.

3.2 Review

In this chapter we have introduced some of the more common terms used when discussing measurement. We need others, most especially when we wish to clarify and quantify uncertainty in measurement. To this end the GUM introduces several new terms, including standard uncertainty, coverage factor and expanded uncertainty. We will define and use these terms in forthcoming chapters.

4

Introduction to uncertainty in measurement

In this chapter we describe how consistency and clarity may be brought to the calculation and expression of uncertainty in measurement.

The goal of any measurement is to establish a numerical value for the measurand. Depending on the accuracy that we wish to claim for the numerical value, the procedure that gives us the value may be relatively simple and direct, involving no more than a tape-measure, for example. In other situations the process may be more complicated, with several intermediate stages requiring the resources of a well-equipped laboratory. Thus, if the measurand is the width of a table, the tape-measure is all that is needed. On the other hand, if the measurand is the accurate mass of an object, we need to know the value of the buoyancy correction (since the weight of the object is less by an amount equal to the weight of the volume of air that it displaces). This in turn requires knowledge of the volume of the object and of the density of air (which is a function of temperature, pressure and composition) at the time of measurement.

There are three components of a measurement: the measurand itself; the measuring instrument (which can be a stand-alone instrument such as a thermometer, or a complex system that occupies a whole laboratory); and the environment (which includes the human operator). The environment will, in general, affect both the measurand and the measuring instrument.

4.1 Measurement and error

4.1.1 Specifying measurand and environment

Error is the difference between the measured value of a measurand and the true value of the measurand. The true value cannot be known; it is an unreachable ideal in an imperfect world. However, we can regard it as the value close to the value that we would obtain if we could specify both the measurand and its environment in

35

very great detail, and if we possessed a measuring instrument of very high accuracy that was traceable to international standards.

If we do not specify the measurand in sufficient detail, then it is not fully defined, and so two people, measuring what they think is the same measurand, may actually make measurements under slightly different conditions, obtaining different values for this reason alone. As an example, the task might be to measure the diameter of a cylindrical brass rod. Here the diameter is the measurand. Although the rod may look cylindrical to the eye, its diameter will actually vary slightly, because of imperfections in the lathe that was used to turn the rod. As part of the process of fully specifying the measurand, we would therefore need to specify *where* the diameter should be measured – say, at the mid-point of the rod.

Similarly, if we do not specify the environment in sufficient detail, we are in effect neglecting the possibility that the measurand may be sensitive to the environment. In the above example, since brass expands or contracts with a rise or drop in temperature, we would need to specify the temperature of the environment in order to specify the measurand. By contrast, we note that, in attempting to estimate the true value of the measurand, we should not have to specify the instrument to be used (for example, its type, manufacturer or model), except for demanding that the instrument should have a certified very high accuracy.

The true value is, then, the value we would obtain for a completely specified measurand if we could use an ideal instrument in a completely specified environment. So we expect an error when we measure the measurand in an imperfect but more practically realisable manner.

Next we recognise that errors come in two flavours: 'random' and 'systematic'.

4.1.2 Random errors

The distinction between random and systematic errors is best seen by considering the notion of 'repeating the measurement under unchanging conditions', or as closely as we can arrange such conditions. By 'unchanging conditions' we mean a well-defined measurand, a tightly controlled environment and the same measuring instrument. Often when we repeat the measurement in this way, we will obtain a different value.[1] The reason for this lack of perfect repeatability is that the instrument we use or the measurand, or both, will be affected by *uncontrollable and small* changes in the environment or within the measurand itself. Such changes may be due, for example, to electrical interference, mechanical vibration or changes in temperature. So if we make the measurement ten times, we are likely to get ten

---

[1] We may obtain exactly the same value simply as a result of the limited *resolution* of the instrument – for example, if a digital instrument displays only two or three digits.

Table 4.1. *Voltage values as displayed by a DMM and difference from the mean voltage of 2.889* μV

DMM indication (μV)	Differences from mean (μV)
2.87	−0.019
2.91	+0.021
2.89	+0.001
2.88	−0.009
2.87	−0.019
2.88	−0.009
2.86	−0.029
2.95	+0.061
2.88	−0.009
2.90	+0.011

values that, although similar, vary by a small amount. When our intention is to obtain a single value for the measurand, we interpret such variations as the effect of errors. The errors fluctuate, otherwise we would see no variation in our values. Errors that fluctuate, because of the variability in our measurements even under what we consider to be the same conditions, are called *random* errors. In brief, random errors arise because of our lack of total control over the environment or measurand.

The first column of table 4.1 contains ten values in microvolts (millionths of a volt, symbol μV) recorded by a digital multimeter (DMM) once a second in a temperature-controlled laboratory. The values were obtained during the calibration of a source of constant voltage of nominal value 1 V. The small values of voltage are the differences in voltage that the DMM indicates between the source and a known and very stable value of a voltage standard. The mean of these ten values is 2.889 μV. The second column shows the ten differences between the measured values and this mean value, which is the measurand.

The differences sum to exactly zero (as all differences from a mean value must do), so both plus and minus signs must be present. These differences are scattered over a 0.090-μV range extending from −0.029 μV to +0.061 μV. This scatter, or 'dispersion', creates an *uncertainty* in the value obtained for the measurand.

This lack of total control over the environment, creating random errors, also affects cases where we make intentional changes to the environment. For example, the electrical resistance of a conducting material varies with temperature. To measure its temperature coefficient of resistance, we measure the resistance at intentionally different temperatures. When the resistance is a very stable and accurately known resistance making up what is known as a 'standard resistor',[2] we require

---

[2] A standard resistor is an example of an artefact standard (see section 3.1.4 and figure 3.2).

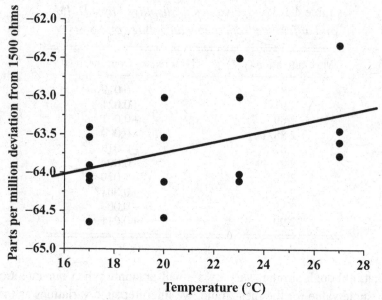

Figure 4.1. Random errors when measuring the temperature coefficient of a resistor (courtesy of the National Measurement Institute of Australia).

the distribution of temperature over its surface to be as uniform as possible. It is
therefore immersed in a tank of stirred oil that can be set to various temperatures.
We cannot *fully* control the temperature, however; nor its distribution over the body
of the resistor. There may also be small fluctuations in the indication of the measuring instrument, possibly because the connecting wires pick up electromagnetic
interference (from power-line and TV transmissions, for example). In brief, the
environment has a basic randomness or 'noise' that we are unable to eliminate
completely. So if we plot the measured resistance against temperature, as in figure
4.1, we are likely to observe a scatter of random errors around the 'line of best fit'
that gives us the temperature coefficient of resistance. In this example, the resistor
has a 'nominal' value of 1500 Ω and is wound from a special type of wire with a
very low temperature coefficient. Several measurements have been taken at each of
four selected temperatures. The temperature coefficient in figure 4.1, namely the
slope of the line of best fit in the figure, is about $+0.071$ $\mu\Omega/\Omega$ $(^{\circ}C)^{-1}$ and, as will
be discussed in section 5.2.3, the scatter of the points about this line can be given
a quantitative value, namely 0.59 $\mu\Omega/\Omega$.

A sequence of reasonably stable measurements suggests a possible general way
in which we might obtain the true value of a measurand. We make as many measurements as possible under the same conditions and calculate their mean. It is often
correct that, in calculating the mean, the random errors will tend to cancel out,
and their cancellation will yield a net error that we can claim with high confidence

to be very nearly zero if there is a very large number of measurements. Making many measurements is in fact generally preferable to making only a few – time and resources permitting. This applies also to cases where intentional changes are made, as in the values shown in figure 4.1; the greater the number of measurements, the more precisely we might expect to establish the value of the temperature coefficient. However, it is often pointless to take *very* many measurements to ascertain the true value of the measurand. The reason is the probable existence of the other flavour of error: a *systematic* error.

4.1.3 Systematic errors

During any measurement, there will probably be an error that remains constant when the measurement is repeated under the same conditions. An example of such an error is a constant offset in a measuring instrument. Unlike random errors, such *systematic* errors cannot be reduced by repeating the measurements and taking their mean; they resist statistical attack. The DMM in the above example (see table 4.1) might consistently – but unknown to us – have an offset, so that it indicates 1 μV too high no matter how many measurements we make. This systematic error will then be transferred to the value of voltage that we finally calculate for the voltage source. On the other hand, we might expect a measurement of temperature coefficient, as in figure 4.1, to be less susceptible to the effect of an offset. As may be checked, a constant offset in the temperature values or resistance values will shift the line in figure 4.1 left or right or up or down, but will not affect its slope.

An instrument may have a systematic error other than an offset. An offset, as commonly understood, is an additive (or subtractive) systematic error, as in the case of the DMM that reads 1 μV too high. A systematic error may also be multiplicative. In the case of a DMM, such an error is often called a 'gain error'; for example, the DMM may read three parts per million too low over a particular range of voltages, so that when it displays (for example) 2.000 000 V, the actual value of voltage is 2.000 006 V. In the case of the temperature-coefficient measurement in figure 4.1, such a multiplicative systematic error will affect the slope.

A systematic error may be revealed by one of two general methods. In the following discussion, we use the term 'device' to refer to either an instrument or an artefact. We may look up previously obtained information on the devices used in a measurement. This information may take the form of specifications by a manufacturer or supplier, or look-up tables of physical constants of materials, and previously reported measurements against higher-accuracy devices. We note especially the latter resource: any device, particularly if used in an accurate measurement, should have been calibrated recently. There are laboratories that perform calibrations and issue a *calibration report* for a specified device. The devices of higher accuracy used in the calibration are themselves calibrated against devices of yet higher accuracy. In this manner, all devices are traceable to the 'top of the food chain' – the international primary standard for the particular quantity. We may call this general class of information 'specific information', since it is specific to the actual measurand that is of immediate concern. Any discrepancy between this specific information and the result of the present measurement suggests that there is a systematic error in the present measurement.

The other method of identifying a systematic error is by changing the experimental set-up. The change may be intentional in order to seek out any systematic error, or may occur for other reasons, with the systematic error being discovered 'by accident' as a result of the change. The change may also take place as a slow natural process, generating an increasing and significant systematic error, which, however, remains unsuspected for a prolonged period. In high-accuracy electrical measurements, the slow deterioration in the insulating property of materials, permitting increasing leakage currents, is such a process.

Here are four examples of intentional change that may uncover a systematic error.

1. In high-accuracy electrical measurements of voltage, swapping the electrical leads connecting a source of constant voltage to a high-accuracy DMM can reveal the systematic errors arising from the DMM's 'zero-offset' and from small thermal voltages caused by the Seebeck effect. The zero-offset error is a non-zero DMM reading when it should be exactly zero (as when a short-circuiting wire is connected across the input terminals), and is due to imperfections in the DMM's internal electronics. The Seebeck effect creates small voltages at junctions between different metals at different temperatures.[3]

2. Exchanging one instrument for another that is capable of the same accuracy and preferably made by a different manufacturer.

3. Having a different person perform the measurement. Thus the exact position of a marker on a scale or of a pointer on a dial will be read differently by different people (a case of so-called 'parallax' error, caused by differences in the positioning of the eye relative to an observed object). In high-accuracy length measurements, using gauge blocks of standard thicknesses, the blocks must often be wrung together to form a stack, and the wringing process, which will determine the overall length of the stack, varies with the operator.

4. An established method of measurement and a novel method that promises higher accuracy may give discrepant results, which will be interpreted as revealing a systematic error in the older method.

An example of such a novel method occurred in high-accuracy measurements of voltage in the early 1970s. Until then, a standard of voltage was provided by banks of

---

[3] These systematic errors are usually no larger than several microvolts when copper wiring is used for the electrical connections. In section 6.2 these errors are discussed in greater detail.

standard cells. These are electrochemical devices containing mercury and cadmium and their sulfates in sulfuric acid, which were developed by Edward Weston in the 1890s, and provide a stable voltage of about 1.018 V at room temperatures (Vinal 1950). However, in 1962 Brian Josephson predicted an effect in superconductors that radically changed the situation.[4] The prediction was that a constant ('direct-current' or 'dc') voltage, V, would exist across a very narrow gap of the order of nanometres (a 'Josephson junction') between two superconductors,[5] if the gap was irradiated by microwaves at frequency f. Crucially for electrical metrology, the relation $V = n[h/(2e)]f$ would be obeyed. The Planck constant, h, and the electron charge, e, are constants of nature, n is a known integer selected by the experimenter, and a frequency, f, can be measured with extremely high accuracy. So this was potentially, and turned out to be in practice, a much superior method of maintaining a standard of voltage compared with the use of standard cells. In many countries the as-maintained unit of voltage was changed as a result of the new method; in the case of Australia, this amounted to a change of about half a part per million introduced in January 1973. Later, in January 1990, all countries that based their voltage standards on the Josephson effect made a further and larger change of about eight parts per million, as a result of absolute measurements of voltage.[6]

In this fourth category we may also include measurements of fundamental constants where there can be no established method and where, because the measurand is a fundamental constant, any variation in results is attributed to experimental error with a strong systematic component. The extremely challenging measurements of the gravitational constant, G, constitute a prime example. Figure 4.2 summarises measurements (Quinn *et al.* 2001) made between 1997 and 2001 of G, which has an approximate value 6.68×10^{-11} m$^3 \cdot$ kg$^{-1} \cdot$ s^{-2}. The horizontal 'error-bars' in figure 4.2, some of which do not overlap, indicate the difficulty of assigning an uncertainty to the measured value of G.

Random and systematic errors have contrasting natures. Random errors can be revealed when we repeat the measurement while trying to keep the conditions constant. Systematic errors can be revealed when we *vary* the conditions, whether

---

[4] Josephson's paper with this discovery is cited, and practical voltage standards based on the Josephson effect are described, in the paper by Hamilton *et al.* cited at the end of chapter 1.

[5] As used in voltage standards, these superconductors are metals (for example, niobium) cooled to temperatures near absolute zero. At low temperature these metals have zero electrical resistance and are therefore known as superconductors.

[6] 'Absolute' electrical measurements, which are invariably complex and demand major laboratory resources, are those made by direct reference to the 'mechanical' standards of mass, length and time. The mercury-electrometer project was such a measurement. It involved defining a voltage through the measurement of the small elevation (a fraction of a millimetre) of a liquid-mercury surface when attracted upwards by a high voltage. The density of mercury and the acceleration due to gravity needed to be accurately known, and a major engineering feat in this experiment was the successful isolation of the system from mechanical vibration (Clothier *et al.* 1989).

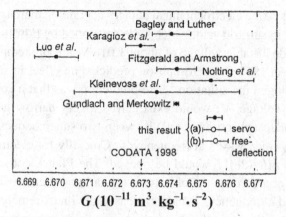

Figure 4.2. Measurements of *G* by various groups between 1997 and 2001 (courtesy T. J. Quinn, National Physical Laboratory, UK).

deliberately or unintentionally. Varying the conditions can be done in a relatively minor way, as in the lead-swapping example above, or it may amount to a major change in the experimental method and system. In general, the bigger the change, the greater the chance of uncovering systematic errors. Deliberately varying the conditions is more troublesome and time-consuming than simply repeating a measurement; this is one reason why systematic errors can remain hidden and unsuspected for prolonged periods.

We see that, even though the existence of *standards of measurement* is fundamental to metrology, *diversity of methods and procedures* is a powerful defence against systematic errors. Indeed, the richness of metrology derives in part from the continuing interplay of these two apparently discordant principles.

Both methods of revealing systematic errors – specific information and changes to the experimental set-up – require a good grasp of the science underlying the measurement. Since any attempt at accurate measurement is potentially or actually beset by systematic errors from many sources – awareness of this is part of the mental atmosphere of metrology – it is useful to have some familiarity with scientific areas apart from the area of immediate relevance to the measurement. As an example, the elaborate experiment mentioned above to measure the 'absolute volt', using a carefully designed mercury electrometer, demanded expertise not only in electricity and magnetism, but also in optics, the physics and chemistry of liquids, metallurgy and other disciplines.

No sharp distinction is to be made between the two ways in which systematic errors are revealed. Specific information can be obtained from the calibration report on an instrument, and the procedure of calibration itself involves a change in the experimental set-up. Nevertheless, the two-way classification serves as a useful reminder of the practical methods by which constant vigilance against systematic errors can be maintained.

After the existence and cause of a systematic error have been established, an experimental routine can often be developed that automatically takes it into account and eliminates it from the final result. Since the magnitude of the systematic error cannot be known exactly, this process of elimination must itself leave an uncertainty. An example of such an experimental routine, which readily lends itself to statistical analysis, concerns the systematic error caused by DMM offset and by a thermal voltage, as mentioned above and as will be described more fully in chapter 6. In other cases, we have to remove the systematic error through an actual calculation that *corrects* for it. For example, if scales consistently overestimate a nominal 1-kilogram mass by 9 grams, then a value of, say, 989 grams should be corrected to 980 grams.

The correction itself is likely to be known only approximately. Errors must therefore be associated with the correction, and we can regard them as random errors scattered around the correction. We note that looking up specific information is hardly a usefully repeatable exercise, and it is generally impracticable to vary the experimental conditions more often than, say, twice. However, just as in the previous case of usefully repeatable measurements with their 'visible' or *explicit* scatter, the *uncertainty* of the correction can be estimated as representing notionally the *implicit* scatter of its associated random errors. So, whether or not we have usefully repeatable measurements, the measurand is measured with an uncertainty that is described as follows.

4.2 Uncertainty is a parameter that characterises the dispersion of values

The dispersion of data is characterised numerically by a standard deviation (defined in section 4.3). From this standard deviation, it is common practice to obtain a '±' figure. This figure describes the range of values that is very likely to enclose the true value of the measurand. The number following the '±' is normally about twice the standard deviation of the measurand and can be loosely referred to as the 'uncertainty' attaching to the measurand. As will be discussed in chapter 10, this uncertainty is referred to in the GUM as the 'expanded' uncertainty, expressing the 'expansion' by that factor of about two from the standard deviation of the measurand.

If a value of a mass is given as (1.24 ± 0.13) kg, the actual value is asserted as very likely to be somewhere between 1.11 kg and 1.37 kg. The uncertainty is 0.13 kg and we note that uncertainty, like standard deviation, is a positive quantity. By contrast, an error may be positive or negative.

4.2.1 Type A and Type B categories of uncertainty

These do not differ in essence, but are given these names in order to convey the notion that *they are evaluated in different ways*.

4.2.1.1 Type A uncertainties are evaluated by statistical methods

In a common situation, a sequence of repeated measurements giving slightly different values (because of random errors) is analysed by calculating the mean and then
considering individual differences from this mean. The scatter of these individual
differences is a rough indication of the uncertainty of the measurement: the greater
the scatter, the more uncertain the measurement.

The calculation of the mean, by summing the values and then dividing this sum
by the number of values, is perhaps the simplest example of statistical analysis. The
scatter around the mean contributes a Type A uncertainty to the uncertainty of the
mean. In a more complicated example requiring statistical analysis, a quantity may
change with time, so that the rate of change, commonly called 'drift', is of interest.
Often this drift is partially or almost completely obscured by random scatter, as in
the case of measurements of climate where a long-term change in temperature may
be masked by day-to-day fluctuations. To tease out the value of temperature drift
from this background 'noise' is a matter for statistical analysis. The numerical value
of drift then has an uncertainty determined by the amount of scatter. This again will
be a Type A uncertainty. In exactly the same way, the value of the temperature
coefficient in figure 4.1 has an uncertainty determined by the scatter (about the
best-fit line) of the 18 measurement points.

4.2.1.2 Type B uncertainties are evaluated by non-statistical methods

A Type B uncertainty may be determined by looking up specific information about
a measurand such as that found in a calibration report or data book. When the
specific information consists of the calibration report on a device, the value of the
measurand is stated in the report – this is the 'calibrated value'. The calibration
report also includes the estimated uncertainty in the value of the measurand. The
calibrated value can tell us how much systematic error would exist if we ignored
the calibration report, and obtaining this information is the primary purpose of
calibrating a device. The uncertainty of the calibrated value is always Type B, from
our point of view as reader and user of the report. The reason is that no statistical
analysis can or needs to be done when reading the report; unlike in the typical case
of Type A uncertainty discussed in section 4.2.1.1, reading the report several times
will give exactly the same result!

The values summarised in the report were presumably obtained from repeated
measurements with an associated Type A uncertainty. The calibration is likely to
have entailed repeated measurements in order to cancel out as much as possible
any random fluctuations and to check the stability of the instrument or artefact.
In the calibration report the measurements are summarised and the uncertainty of
the result is estimated using statistical methods. This uncertainty will therefore
have a Type A component. Suppose that – unrealistically but as an illustration –

the measurements made by the calibrating laboratory have no Type B component of uncertainty. The uncertainty, which was wholly Type A as determined by the calibrating laboratory, is a Type B uncertainty from the point of view of the reader of the report. The act of writing the report 'fossilises' a Type A uncertainty into a Type B uncertainty. If the reader of the report now uses the reported value in some subsequent application of the instrument or artefact, the uncertainty that was stated in the report is Type B.

When the specific information consists of manufacturer's specifications, the contents of tables of physical properties or the like, it often happens that no associated uncertainties are stated. The information provided by these sources will remove the systematic error that would be present if we used only an approximate value. However, we then have to estimate the associated uncertainty ourselves, without benefit of either statistical analysis or a reported uncertainty. As discussed in section 2.3, this Type B uncertainty can often be estimated from the stated number of decimal places.

4.2.2 Combining Type A and Type B uncertainties

Specific information or changing the conditions of an experiment, whether deliberately or accidentally, may reveal an unsuspected systematic error. This error must itself have an associated uncertainty. After the error has been corrected for, this uncertainty may be Type A or Type B and is then combined with the Type A uncertainty evaluated from random errors. Depending on the particular circumstances, both the Type A and the Type B uncertainties may or may not be reported separately. However, what is always reported is the uncertainty formed from the combination of the Type A and Type B components. From the point of view of the user of the report, this combined uncertainty is wholly Type B.

Figure 4.3 illustrates the relationships among the errors and uncertainties.

4.3 Standard deviation as a basic measure of uncertainty

If there are n values of a quantity, $x_1, x_2, \ldots, x_n$, the standard deviation, s, of these n values is given by[7]

$$s = \sqrt{\frac{\sum_{i=1}^{n}(x_i - \bar{x})^2}{n-1}}, \tag{4.1}$$

where $\bar{x}$ is the mean of the n measurements, defined as $\bar{x} = (1/n)\sum_{i=1}^{n} x_i$.

---

[7] Strictly this is an approximate estimate of the standard deviation. This is considered in more detail in sections 5.1.3 and 5.1.4.

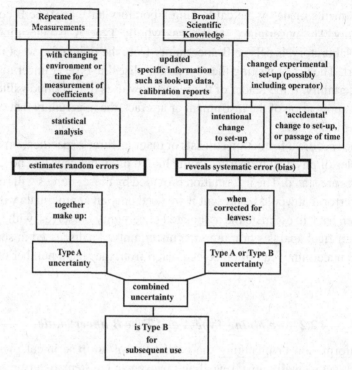

Figure 4.3. The relationship between Type A and Type B uncertainties.

The value of s in the example of small voltage differences in table 4.1 is about 0.026 µV. We note that this is substantially less than the overall range (0.090 µV) by a factor of roughly 3.5. The standard deviation is often less than the overall range of the values (or of the random errors) by a factor between 3 and 4.

The square of s, s^2, is known as the *unbiased variance* of the x_i ($i = 1, 2, \ldots, n$), or more exactly the unbiased estimate of the variance of the entire population of the x's of which our n values form a sample. The variance s^2 of the population is, then,[8]

$$s^2 = \frac{\sum_{i=1}^{n}(x_i - \bar{x})^2}{n - 1}. \tag{4.2}$$

The spread of values is a source of uncertainty in the final result. Since the standard deviation is a measure of the spread, the name given in metrology to the standard deviation is 'standard uncertainty'. The symbol frequently used for standard uncertainty is a lower-case u, so that $u(x)$ is the standard uncertainty of a quantity x. Similarly, $u^2(x)$ denotes the variance of x.

---

[8] Section 5.1.3 discusses the reason for the presence of $n - 1$ rather than n in the denominator of equation (4.2).

Table 4.2. *Number of airborne
particles in a fixed volume
in a cleanroom*

Number of particles
137
114
88
102
95
102

Example 1

Six successive measurements of the number of airborne particles within a fixed volume of air within a clean room are made. Table 4.2 show the values obtained. Use these data to calculate (a) the variance and (b) the standard uncertainty in the number of particles.

Answer

(a) Calculation of the variance using equation (4.2) is best accomplished using an electronic calculator or a spreadsheet package, such as Excel. Such a calculation gives $s^2 = 300.3 = u^2(x)$, where x represents the number of particles.

(b) Since $u^2(x) = 300.3$, the standard uncertainty in the number of particles is $u(x) = \sqrt{300.3} = 17.3$.

Exercise A

(1) Ten samples of an oxide of nominally the same mass are heated in an oxygen-rich atmosphere for 1 hour. The mass of each sample increases by an amount shown in table 4.3. Using the data in table 4.3, calculate the variance and the standard uncertainty of the mass gain.

(2) The thickness of an aluminium film deposited onto a glass slide is measured using a profilometer. The values obtained from six replicate measurements are shown in table 4.4. Using these data, calculate the variance and standard uncertainty in the film thickness.

The standard uncertainty, $u(\bar{x})$, of the mean $\bar{x} = (1/n) \sum_{i=1}^{n} x_i$ may be expected to be less than s. This is correct if the values x_i ($i = 1, 2, \ldots, n$) are uncorrelated.[9] If they are uncorrelated, then

$$u(\bar{x}) = s/\sqrt{n} \qquad \text{for uncorrelated measurements.} \qquad (4.3)$$

[9] This is discussed more fully in chapters 5 and 7.

Table 4.3. *Mass gained
by samples of oxide*

Mass gain (mg)
12.5
11.2
11.8
11.8
12.1
11.5
11.0
12.1
11.7
12.8

Table 4.4. *Thickness of
aluminium film*

Thickness (nm)
420
460
400
390
410
460

In the case of the voltage differences in table 4.1, where $s = 0.026$ μV, the overall correlation among the ten measurements may be shown to be low and the value of $u(\bar{x})$ is $0.026/\sqrt{10}$ μV ~ 0.008 μV.

$u(\bar{x})$ is sometimes called the experimental standard deviation of the mean (ESDM). (In some books this is referred to as the 'standard error' of the mean.) The ESDM when obtained using equation (4.3), with a divisor $\sqrt{n}$, should be used with caution. Equation (4.3) is valid only for uncorrelated values; if, for example, the values exhibit a steady drift in time, then this high correlation implies that the ESDM is not significantly less than s and in fact is closely equal to it. This topic will be discussed further in section 7.2.2.

Example 2
Calculate the mean, $\bar{x}$, and standard uncertainty in the mean, $u(\bar{x})$, for the values in table 4.2.

Answer

The mean of the values of the number of airborne particles in table 4.2 is $\bar{x} = 106.3$. $u(\bar{x})$ is given by equation (4.3). Here, $s = 17.3$ and the number of values is $n = 6$, so $u(\bar{x}) = 17.3/\sqrt{6} \simeq 7.1$.

Exercise B

(a) Using the data in table 4.3, calculate the mean mass gain and standard uncertainty in the mean.

(b) Using the data in table 4.4, calculate the mean thickness of the aluminium film and the standard uncertainty in the mean.

We can see intuitively why, when the values are uncorrelated, the standard uncertainty, $u(\bar{x})$, of the mean is less than the standard deviation, s, of the scatter. When the mean is calculated, the random errors tend to cancel out. This follows from the fact that the measured values are summed for calculating the mean, and the random errors come with both positive and negative signs.[10] However, the cancellation is itself an uncertain process; this is why the reduction from s to $u(\bar{x})$ is not by a factor of n, but only by a factor of $\sqrt{n}$. So, if we go to the trouble of taking not 10 but 100 uncorrelated measurements, the standard uncertainty of the mean will be reduced further only by a factor of about three.

4.4 The uncertainty in the estimate of uncertainty

If the standard uncertainty is denoted by s and its own uncertainty by $u(s)$, then[11]

$$u(s) \sim \frac{s}{\sqrt{2\nu}}. \tag{4.4}$$

where ν is the number of 'degrees of freedom'.[12] ν is equal to the number of values, n, minus the number of quantities determined using the values. In the case where the mean is the only quantity determined using the values, $\nu = n - 1$.

Expressed as a percentage uncertainty, we can write equation (4.4) as

$$\frac{u(s)}{s} \times 100\% \sim \frac{1}{\sqrt{2\nu}} \times 100\%. \tag{4.5}$$

If ν is as low as 4, then $u(s)/s$ is high at about 35%, and ν has to reach 50 to give a 10% uncertainty in the uncertainty.

Equation (4.5) is particularly useful for the Type B category of uncertainty. Owing to the tentative nature of the estimation of Type B uncertainties, it is good

---

[10] As shown in the voltage-measurement example in section 4.1.2.
[11] Equation (4.4) is discussed further in section 9.3.1.
[12] We will consider degrees of freedom more fully in section 5.1.5.

Table 4.5. *Minimum force*
required to move a glass block

Force (N)
5.6
5.7
5.2
5.5
5.8
5.7
5.4

to have some numerical indicator of the reliability which we think attaches to a Type B estimate. Often the value of v for a Type B uncertainty will not exceed 10, implying a reliability in the estimated uncertainty of no better than about 20%.

Equation (4.5) may be used to determine the uncertainty in a Type A uncertainty. For example, for the data in table 4.1, $s = 0.026$ μV, and $v = n - 1 = 9$. On substituting these values into equation (4.5) we find that the percentage uncertainty in s is surprisingly high at almost 25%. The percentage uncertainty in the standard uncertainty in the mean, $u(\bar{x})$, is also 25%. Since $u(\bar{x}) = 0.008$ μV, the uncertainty in $u(\bar{x})$ is about 0.002 μV.

Exercise C

Table 4.5 shows repeat measurements of the minimum force required to cause a glass block to move when it is resting on a smooth metal plate. Using these data, determine

(a) the mean minimum force to move the glass block,
(b) the standard uncertainty in the mean and
(c) the uncertainty in the standard uncertainty.

4.5 Combining standard uncertainties

A measurand may be measured indirectly, through the measurement of so-called 'input quantities'. If y is the measurand and $x_1, x_2, \ldots, x_n$ are the input quantities, then y is a function $y = f(x_1, x_2, \ldots, x_n)$ of the x's. The standard uncertainty, $u(y)$, in y resulting from standard uncertainties $u(x_1), u(x_2), \ldots, u(x_n)$ in the input quantities is calculated using the equation

$$u^2(y) = \left(\frac{\partial y}{\partial x_1}\right)^2 u^2(x_1) + \left(\frac{\partial y}{\partial x_2}\right)^2 u^2(x_2) + \left(\frac{\partial y}{\partial x_3}\right)^2 u^2(x_3) + \cdots + \left(\frac{\partial y}{\partial x_n}\right)^2 u^2(x_n).$$

$$(4.6)$$

Equation (4.6) may be written more compactly:

$$u^2(y) = \sum_{i=1}^{n} \left(\frac{\partial y}{\partial x_i}\right)^2 u^2(x_i). \tag{4.7}$$

Equation (4.6) is valid *only* if the x_i are mutually uncorrelated. A correlation exists if, for example, two or more of the x_i have been measured using the same instrument that has a systematic error with a significant associated uncertainty. The generalisation of equation (4.6) for the correlated case is discussed in section 7.2.

The standard uncertainties of the inputs, namely the u's on the right-hand side of equation (4.6), may be either Type A or Type B uncertainties. *No distinction between Type A and B is made when evaluating the standard uncertainty of the measurand, y.*

Example 3
The velocity of a wave, v, is written in terms of the frequency, f, and the wavelength, λ, as

$$v = f\lambda. \tag{4.8}$$

An ultrasonic wave has $f = 40.5$ kHz with a standard uncertainty of 0.15 kHz and $\lambda = 0.822$ cm with a standard uncertainty of 0.022 cm. Assuming that there is no correlation between errors in f and λ, calculate the velocity of the wave and its standard uncertainty.

Answer
The velocity $v = f\lambda$ is given by $v = (40.5 \times 10^3) \times (0.822 \times 10^{-2}) = 332.9$ m/s.

Writing equation (4.6) in terms of v, f and λ gives

$$u^2(v) = \left(\frac{\partial v}{\partial f}\right)^2 u^2(f) + \left(\frac{\partial v}{\partial \lambda}\right)^2 u^2(\lambda). \tag{4.9}$$

Using equation (4.8), we have

$$\frac{\partial v}{\partial f} = \lambda, \qquad \frac{\partial v}{\partial \lambda} = f. \tag{4.10}$$

Substituting in values gives

$$u^2(v) = (0.822 \times 10^{-2})^2 \times (0.15 \times 10^3)^2 + (40.5 \times 10^3)^2 \times (0.022 \times 10^{-2})^2 \text{ (m/s)}^2, \tag{4.11}$$

hence $u^2(v) = 80.9$ (m/s)^2, so that $u(v) = 9.0$ m/s.

Exercise D

(1) The flow rate of blood, Q, through an aorta is found to be 81.5 cm^3/s with a standard uncertainty of 1.5 cm^3/s. The cross-sectional area, A, of the aorta is 2.10 cm^2 with a standard uncertainty of 0.10 cm^2. Find the flow speed of the blood, v, and the standard uncertainty in the flow speed using the relationship[13]

$$Q = Av. \tag{4.12}$$

(2) The velocity, v, of a wave on a stretched string is given by

$$v = \sqrt{\frac{F}{\mu}}, \tag{4.13}$$

where F is the tension in the string and μ is the mass per unit length of the string. Given that $F = 18.5$ N with a standard uncertainty of 0.8 N and $\mu = 0.053$ kg/m with a standard uncertainty of 0.007 kg/m, calculate the velocity of the wave and its standard uncertainty.

Historical note. It used to be the common practice, before the introduction of the GUM, for measurement and testing laboratories to report uncertainties as so-called 'errors'. It was also common to report separately the random and systematic errors in the measurand. This often created the complication that, in any subsequent use of the report by others, a single number for the uncertainty, though desirable, was not immediately apparent. There was no consensus regarding the measure of uncertainty: whether this should be the standard deviation or a small multiple of this. Instead of the root-sum-square rule, errors and/or uncertainties were often simply summed linearly. This linear sum applies strictly to perfectly positively correlated input quantities, and if there is little or no correlation the linear sum gives a needlessly pessimistic estimate of the uncertainty in the measurand.

4.6 Review

While errors are conveniently categorised as random or systematic, the GUM introduces the new terms 'Type A' and 'Type B' to categorise uncertainties. Type A and Type B uncertainties are not related directly to random and systematic errors, but reflect the way in which uncertainties are evaluated. In the next chapter we will turn our attention to useful statistical methods that allow us to summarise key features of experimental data.

---

[13] Equation (4.12) is often referred to as the 'continuity equation'.

5

Some statistical concepts

Random errors arise from uncontrollable small changes in the measurand, instrumentation or environment. These changes are evident as variations in the values obtained when we carry out repeat measurements. In this chapter we shall consider methods of quantifying these variations: that is, describing them numerically using statistical methods. Some basic statistical concepts will therefore be introduced and discussed.

5.1 Sampling from a population

In statistics, the term *population* refers to the number of *possible*, but not necessarily *actual*, measured values. In some situations a population consists of an infinite number of values. In practice, we can measure only a sample drawn from a population, since time and resources are always limited. We hope and expect that the sample is representative of the population. In almost every case of measurement we sample a population, and the quantities of interest obtained from the sample (sometimes called *sample statistics*) should reliably represent corresponding parameters in the population (the *population parameters*). An example of such a quantity of interest, which quantifies the amount of scatter in values, is the standard deviation of the values.

There are cases where a sample may, in fact, be the entire population. Thus the examination results of a class of 30 students can be analysed statistically in order to determine, for example, the mean mark and the range of marks, with no attempt at generalising. The teacher of the class may be interested simply in that particular class. But normally, when measurements are made, a sample is implicitly understood to be representative of the underlying population; if it were not, the measurements made by a particular person in a particular laboratory would be of little interest to anybody else!

5.1.1 *The expectation value of a continuous random variable*

Most often we routinely measure quantities, such as temperature and time, that are *continuous*, by which we mean that the quantities can take on any value between physically prescribed limits. It is convenient mathematically to represent such a quantity by a variable like x or z (often referred to as a *continuous random variable*). We define the expectation, $E(z)$, of a variable, z, as the mean of that variable over its entire population. The expectation function has the following properties.[1]

(a) The expectation of a constant is that constant. If C is a constant, then $E(C) = C$.
(b) The expectation of the product of a constant and a variable is the product of the constant and the expectation of the variable: $E(Cz) = CE(z)$.
(c) The expectation of the sum of variables z_1, z_2, ... is the sum of the expectations of those variables: $E(z_1 + z_2 + \cdots) = E(z_1) + E(z_2) + \cdots$.

If we give the constant in (b) the value -1, then (c) implies that the expectation of the difference between two variables is the difference between their expectations: $E(z_1 - z_2) = E(z_1) - E(z_2)$.

(d) The expectation of the product of variables z_1, z_2, ..., z_n is the product of the expectations if, and only if, the variables z_1, z_2, ..., z_n are mutually uncorrelated.[2] Then we have $E(z_1 z_2 \ldots z_n) = E(z_1)E(z_2) \ldots E(z_n)$.

If the expectation of a sample statistic is the value of the corresponding population parameter, the sample statistic is said to be an *unbiased* estimate of the population parameter.

5.1.2 *The mean of a sample and the mean of the population*

Let n denote the sample size. If x_i ($i = 1, 2, \ldots, n$) are the measured values that make up the sample, the mean, $\bar{x}$, of the sample is given by

$$\bar{x} = \frac{\sum_{i=1}^{n} x_i}{n}. \tag{5.1}$$

It is conventional to denote the population mean by the symbol μ. From the definition of expectation, we have, for all x_i ($i = 1, 2, \ldots, n$),

$$E(x_i) = \mu. \tag{5.2}$$

Equation (5.2) may be understood as follows: for *each* x_i in the sample ($i = 1, 2, \ldots, n$), the expectation of that particular x_i is the same as the expectation

---

[1] For more information on expectation, see Devore (2003).
[2] Correlation will be discussed in section 5.3.

of any other x_j in the sample ($j = 1, 2, \ldots, n$), and each is equal to μ. Therefore there is no suffix i on the right-hand side of equation (5.2).

It is instructive to find the expectation of the mean, $\bar{x}$, of the sample. From equation (5.2) and properties (b) and (c) above of the expectation function, we have

$$E(\bar{x}) = E\left(\frac{\sum_{i=1}^{n} x_i}{n}\right) = \frac{E\left(\sum_{i=1}^{n} x_i\right)}{n} = \frac{\sum_{i=1}^{n} E(x_i)}{n} = \frac{\sum_{i=1}^{n} \mu}{n} = \frac{n\mu}{n} = \mu.$$
(5.3)

Equation (5.3) indicates that the expectation of the sample mean is the population mean. The sample mean, calculated using equation (5.1), is therefore an unbiased estimate of the population mean.

5.1.3 The variance of a sample and the variance of the population

Another important property of a sample is the *range*, where range = (maximum value − minimum value). The range is roughly proportional to, and often about three to four times larger than, the standard deviation, which will be discussed in the next section. It is convenient from a theoretical point of view to consider first the square of the standard deviation, known as the *variance*.

We denote the variance of a sample by s_b^2, defined as

$$s_b^2 = \frac{\sum_{i=1}^{n} (x_i - \bar{x})^2}{n}.$$
(5.4)

(The squared symbol, s_b^2, anticipates the definition of the standard deviation, s_b, and the subscript, b, indicates that s_b^2 will be a biased estimate of the population variance.)

If the population size is represented by N, the population variance, denoted by σ^2, may be written as

$$\sigma^2 = \frac{\sum_{i=1}^{N} (x_i - \mu)^2}{N}$$
(5.5)

and equation (5.5) is equivalent to

$$\sigma^2 = E[(x_i - \mu)^2]$$
(5.6)

since the expectation of any quantity is the mean of that quantity over the whole population.

It may be shown that[3]

$$E(s_b^2) = \sigma^2 \frac{n-1}{n}.$$
(5.7)

[3] See, for example, chapter 8 in Wilks (1962).

Since $E(s_b^2) \neq \sigma^2$, s_b^2 is said to be a biased estimate of σ^2. However, if we define a new quantity, s^2, as

$$s^2 = \frac{n}{n-1} s_b^2 = \frac{\sum_{i=1}^{n}(x_i - \bar{x})^2}{n-1} \tag{5.8}$$

then

$$E(s^2) = [n/(n-1)]E\left(s_b^2\right) = [n/(n-1)][(n-1)/n]\sigma^2 = \sigma^2. \tag{5.9}$$

So the variance defined in equation (5.8), with the divisor $n - 1$, is the unbiased, and therefore preferred,[4] estimate of σ^2.

We note that s^2 is slightly greater than s_b^2 (although, for large sample sizes, the difference between s^2 and s_b^2 is small). This can intuitively be seen to be reasonable. In equation (5.4), s_b^2 is defined as the mean of squared differences between the sample values, x_i ($i = 1, 2, \ldots, n$) and the sample mean, $\bar{x}$. However, $\bar{x}$ is not a fixed quantity; it varies from sample to sample, in contrast to the fixed population mean, μ. Also, $\bar{x}$ is positively correlated with the values x_i (not surprisingly, since $\bar{x}$ is their mean![5]) and so the deviations $x_i - \bar{x}$ are slightly shrunken measures of the range of variability of the values x_i. Dividing by the slightly smaller number $n - 1$, rather than by n, exactly compensates for this shrinking.

If we expand the right-hand side of equation (5.6), we obtain

$$\sigma^2 = E\left(x_i^2 + \mu^2 - 2\mu x_i\right) = E\left(x_i^2\right) + E(\mu^2) - 2\mu E(x_i). \tag{5.10}$$

Since μ is a constant (the population mean), $E(\mu^2) = \mu^2$; and since $E(x_i) = \mu$, equation (5.10) gives

$$\sigma^2 = E\left(x_i^2\right) + \mu^2 - 2\mu^2 = E\left(x_i^2\right) - \mu^2. \tag{5.11}$$

The term $E(x_i^2)$ is the mean of the squares of the measured values x_i. Equation (5.11) therefore states that the population variance may be regarded as the *mean square of the values minus the square of the mean of the values*.[6] The mean square of a set of numbers is always equal to or greater than their squared mean, so the variance is always zero or a positive quantity.

---

[4] Although the divisor in equation (5.8) is $n - 1$, the summation is over all n terms as indicated in the numerator of equation (5.8).

[5] The smaller the sample size n, the larger the positive correlation between $\bar{x}$ and any x_i ($i = 1, 2, \ldots, n$).

[6] When the values are very similar except for several least-significant digits, this simple formula may give serious round-off errors. It is better in such cases to take differences from the mean of the set. For example, if the set consists of the three similar values 1000.013, 1000.021 and 1000.002 with mean 1000.012, the differences from the mean are +0.001, +0.009 and −0.010 (summing to zero, as a check). The variance of these three differences is the same as the variance of the three original values, but the simple formula can now safely be used on the differences. Alternatively, the common amount '1000' can be subtracted from each value, leaving +0.013, +0.021 and +0.002, and the simple formula can equally safely be used on these three values.

Table 5.1. *Values of pH*
of river water

pH
6.8
7.3
6.9
6.9
7.2
7.0
7.1

5.1.4 The standard deviation of a sample and the standard deviation of the population

The standard deviation, s_b, of a sample is the square root of equation (5.4):

$$s_b = \sqrt{\frac{\sum_{i=1}^n (x_i - \bar{x})^2}{n}}. \tag{5.12}$$

We can now define the estimated standard deviation, s, of the population as the square root of the unbiased estimate, s^2, of the population variance, σ^2:

$$s = \sqrt{\frac{\sum_{i=1}^n (x_i - \bar{x})^2}{n - 1}}. \tag{5.13}$$

This is only an approximately unbiased estimate of the population standard deviation. Although $E(s^2) = \sigma^2$ as in equation (5.9), it does *not* follow that $E(s) = \sigma$. However, the standard deviation as defined in equation (5.13) is the accepted measure of the amount of variation of some quantity in its population, as inferred from a sample.

We note that (as is shown by, for example, equations (5.5), (5.8) and (5.13) and implied by footnote 6), variance and standard deviation, whether of a sample or of a population, are *location-independent*; that is, the variance and standard deviation of a set of values remain unaffected if we add the same arbitrary constant to each value. Such location-independence is an essential attribute of any reasonable measure of variability of values. On the other hand, if each value is *multiplied* by an arbitrary constant, the standard deviation will be multiplied by the same constant, and the variance by the square of that constant.

Exercise A
1. As part of a study on the quality of river water, seven repeat measurements were made of the pH of the water. Table 5.1 shows the values obtained. Determine s_b and s using equations (5.12) and (5.13).

2. (a) Writing the percentage difference between s_b and s as

$$\%\text{diff} = \left(\frac{s - s_b}{s}\right) \times 100\% \tag{5.14}$$

show, using equations (5.12) and (5.13), that

$$\%\text{diff} = \left(1 - \sqrt{\frac{n-1}{n}}\right) \times 100\%. \tag{5.15}$$

(b) For what value of n (rounded up to the nearest whole number) does the percentage difference given by equation (5.15) equal (i) 20%, (ii) 5% and (iii) 1%?

5.1.5 Residuals and degrees of freedom

Another way of looking at equations (5.8) and (5.13), with more general applicability, is in terms of so-called 'residuals'. Suppose that we calculate the mean, $\bar{x}$, of n values x_i, $i = 1, 2, \ldots, n$. Using the mean, we calculate the n resulting residuals, ϵ_i ($i = 1, 2, \ldots, n$), where

$$\epsilon_i = x_i - \bar{x} \qquad (i = 1, 2, \ldots, n). \tag{5.16}$$

In general, for a sample of size n,

$$\sum_{i=1}^{n} \epsilon_i = 0. \tag{5.17}$$

To show that equation (5.17) must always hold, we sum equation (5.16) over all values of i from $i = 1$ to $i = n$:

$$\sum_{i=1}^{n} \epsilon_i = \sum_{i=1}^{n} x_i - \sum_{i=1}^{n} \bar{x}. \tag{5.18}$$

$\bar{x}$ does not contain the index, i, so

$$\sum_{i=1}^{n} \bar{x} = \bar{x} + \bar{x} + \cdots + \bar{x} = n\bar{x}. \tag{5.19}$$

Hence equation (5.18) may be written

$$\sum_{i=1}^{n} \epsilon_i = \sum_{i=1}^{n} x_i - n\bar{x} \tag{5.20}$$

and substituting equation (5.1) into equation (5.20) gives equation (5.17).

Since the residuals are linked through equation (5.17), the residuals are not independent. If we are given the values of any $n - 1$ of the residuals, then the value of the remaining one is determined by equation (5.17). This is why we say that the n residuals have degrees of freedom, ν, given by

$$\nu = n - 1. \tag{5.21}$$

We next note that the standard deviation in equation (5.13) may be written as

$$s = \sqrt{\frac{\sum_{i=1}^{n} \epsilon_i^2}{n - 1}} = \sqrt{\frac{\sum_{i=1}^{n} \epsilon_i^2}{\nu}}, \tag{5.22}$$

and the variance, s^2, is given by

$$s^2 = \frac{\sum_{i=1}^{n} \epsilon_i^2}{\nu}. \tag{5.23}$$

In the next section we discuss least-squares fitting, of which the estimation of a population mean is the simplest example, and we shall see how residuals and measures of their scatter such as equations (5.22) and (5.23) occur more generally.

5.2 The least-squares model and least-squares fitting

5.2.1 The mean as a least-squares fit

After a sequence of measurements, we often need to estimate one or more parameters that characterise or summarise the values obtained. Often only one parameter is sought. Using least-squares estimation, this parameter turns out to be the 'average' or *mean*, as will now be shown. This is the simplest example of a least-squares fit. The procedure of least-squares fitting is very commonly used when parameters (such as the mean) are to be estimated.

The *least-squares model* that forms the basis of this procedure is, very generally, as follows:

measured values = (function of one or more parameters to be estimated) + scatter. (5.24)

The right-hand side of equation (5.24) may also include known constants or variables that are assumed to be error-free. These variables are sometimes known as 'explanatory' or 'predictor' variables, and some examples will be given in section 5.2.3. The 'scatter' in equation (5.24) describes the random variations or fluctuations that are generally present when measurements are made. They are distinguished conceptually from the parameters that we are fitting to the measured values. In electronic engineering, equation (5.24) is often reformulated as data = signal + noise.

We now consider a numerical example of a case of least-squares fitting. Suppose that we have collected six small pieces of fruit dropped by a palm tree. We weigh them and find the values in grams: 4.1, 4.3, 4.4, 4.2, 4.3, 3.9.

These values, constituting our sample, are to be summarised by a single statistic that characterises the whole population of the fruit from the palm tree (so we have to be careful that our sample is representative). We know in advance that this statistic is the mean of the six values, namely 4.20 g. But suppose that we were unsure how to summarise the six values, and decided to use least-squares to obtain a best estimate.

We denote the single summarising statistic by m. We therefore have

$$
\begin{aligned}
4.1 &= m + \epsilon_1, \\
4.3 &= m + \epsilon_2, \\
4.4 &= m + \epsilon_3, \\
4.2 &= m + \epsilon_4, \\
4.3 &= m + \epsilon_5, \\
3.9 &= m + \epsilon_6,
\end{aligned}
\tag{5.25}
$$

where ϵ_1, ϵ_2 etc. are measures of the 'noise'.[7] In the least-squares model, a common term for the noise is 'residuals', the quantities 'left over' after we have fitted our parameter or parameters.

There are six equations in equation (5.25), but seven unknowns: m and the six ϵ's. Since there are more unknowns than equations, we cannot find a unique solution unless some extra condition is imposed. This is the least-squares condition, and we impose it in the following way.

We sum the squares of the residuals, calling this sum, Q, where[8]

$$
Q = \sum_{i=1}^{6} \epsilon_i^2 = \epsilon_1^2 + \epsilon_2^2 + \epsilon_3^2 + \epsilon_4^2 + \epsilon_5^2 + \epsilon_6^2.
\tag{5.26}
$$

Using equation (5.25), we may write equation (5.26) as

$$
\begin{aligned}
Q &= (4.1 - m)^2 + (4.3 - m)^2 + (4.4 - m)^2 + (4.2 - m)^2 \\
&\quad + (4.3 - m)^2 + (3.9 - m)^2.
\end{aligned}
\tag{5.27}
$$

The required value of m in equation (5.27) is the value of m that makes Q a minimum; hence the name 'least-squares'. This procedure therefore selects that value of our unknown parameter m that minimises the sum of the squares of the residuals (at this point the residuals have unknown values).

---

[7] The noise has units of g (grams) in this example.

[8] Q has units of g^2 (gram-squared).

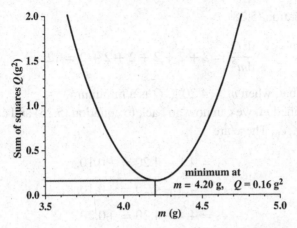

Figure 5.1. The sum of squares, Q, as a function of m as given by equation (5.27).

Equation (5.27) gives Q as a function of m, and this function is a parabola, as illustrated in figure 5.1.

What is the value of m for which Q is a minimum? The gradient of the graph at that point must be zero (the tangent is horizontal), so the derivative of Q with respect to m must be zero at that point. Differentiating Q with respect to m, using equation (5.27), gives

$$\frac{dQ}{dm} = -2(4.1 - m) - 2(4.3 - m) - 2(4.4 - m) - 2(4.2 - m)$$
$$- 2(4.3 - m) - 2(3.9 - m) \tag{5.28}$$

and, since this is equal to zero when Q is a minimum, we have

$$-2(4.1-m)-2(4.3-m)-2(4.4-m)-2(4.2-m)-2(4.3-m)-2(3.9-m) = 0. \tag{5.29}$$

The -2's cancel, giving

$$4.1 + 4.3 + 4.4 + 4.2 + 4.3 + 3.9 - 6m = 0. \tag{5.30}$$

Therefore,

$$m = \frac{4.1 + 4.3 + 4.4 + 4.2 + 4.3 + 3.9}{6} = 4.20. \tag{5.31}$$

The required value of m is the mean of the six results. Substituting this value of m into equation (5.27) gives $Q = 0.16 \, g^2$, the minimum value of Q. Figure 5.1 shows that, at $m = 4.20 \, g$, $Q = 0.16 \, g^2$ and is a minimum.

For a minimum value, the second derivative must be positive (so, as we approach the minimum point from left to right, the first derivative (the gradient) becomes continuously more positive). Differentiating Q with respect to m a second time gives, from equation (5.28),

$$\frac{d^2 Q}{dm^2} = 2 + 2 + 2 + 2 + 2 + 2 = 12. \tag{5.32}$$

This confirms that, when $m = 4.20$ g, Q is a minimum.[9]

Having obtained m, we can now go back to equation (5.25) and calculate the six residuals, $\epsilon_1, \ldots, \epsilon_6$. These are

$$\epsilon_1 = 4.1 - 4.20 = -0.10,$$

$$\epsilon_2 = 4.3 - 4.20 = +0.10,$$

$$\epsilon_3 = 4.4 - 4.20 = +0.20, \tag{5.33}$$

$$\epsilon_4 = 4.2 - 4.20 = 0.00,$$

$$\epsilon_5 = 4.3 - 4.20 = +0.10,$$

$$\epsilon_6 = 3.9 - 4.20 = -0.30.$$

As noted previously, the six residuals must sum to zero and therefore have only five degrees of freedom: $\nu = 5$. We can now calculate the unbiased estimate of the variance, s^2, of the population from which this sample of residuals is drawn, using equation (5.23):

$$s^2 = \frac{\sum_{i=1}^{n} \epsilon_i^2}{\nu}$$

$$= \frac{(-0.10)^2 + (+0.10)^2 + (+0.20)^2 + (0.00)^2 + (+0.10)^2 + (-0.30)^2}{5}$$

$$= \frac{0.16}{5} = 0.032, \tag{5.34}$$

so the standard deviation is $s = \sqrt{0.032}$ g $= 0.18$ g, roughly one-third of the range of values (range $= 4.4$ g $- 3.9$ g $= 0.5$ g).

This standard deviation of the residuals is exactly the same as the estimated standard deviation of the population calculated using equation (5.13). In fact equation

---

[9] What is 'magic' about least-squares – why not 'least-fourth-powers', for example? ('Least-cubes' is immediately unacceptable, since the sum of cubes of residuals could be zero even with large positive and large negative residuals). Least-squares is particularly appropriate in the generally assumed case where the residuals have an approximately bell-shaped or Gaussian distribution (see chapter 8). This is because, in the Gaussian case, least-squares is equivalent to another fitting criterion called 'maximum-likelihood'. Moreover if, in equation (5.27), Q were the sum of fourth powers instead of squares, we would get, instead of $m = 4.20$, the plausible-looking value $m = 4.1631$. However, unlike the mean, this is a biased estimate of the parameter with which we intend to summarise the population. It may further be shown that, amongst all linear unbiased estimates of this parameter, least-squares gives the estimate that is most stable against fluctuations in the original values (the estimate is a 'minimum-variance estimate'). See Seber (1977) for more information.

(5.13) gives the standard deviation, s, in terms of differences from the mean, namely in terms of the residuals indicated in equation (5.16). The definition of s as given by equation (5.13) is equivalent to equation (5.23), where the residuals appear explicitly. By way of contrast, we shall now consider another instance of least-squares fitting in which the standard deviation of the residuals can be significantly less than the standard deviation of the original values.

5.2.2 Patterns obtained in repeated readings

Repeated readings made over time may reveal an overall pattern or trend even though random errors are present. The experimenter must decide whether or not there is an overall pattern and, if there is, whether or not it must be taken into account. Deciding whether there is an overall pattern may be done 'by eye', as when the results are plotted on a graph and the plot is then inspected for any steady change or drift (for example). Testing for any overall pattern can also be done using statistical methods. A pattern may be obvious over a short run of readings, but disappear or lessen in significance over a longer run. If a pattern exists, the second decision is then made as to whether it has any importance. This will depend on the amount of information expected to be gleaned from a particular investigation; thus a very obvious drift may be ignored if all that is required is the mean of the readings or their overall range.

In the case that we illustrated in figure 4.1 of the temperature-coefficient measurements on a standard resistor, the pattern is a general rise in the measured values as the temperature is increased. There is a large scatter, but a general rise can be discerned 'with the naked eye', and we are entitled to conclude that the resistance increases with temperature; there is a temperature coefficient of resistance and it is positive.

Figure 5.2 shows a plot of 30 readings of air temperature in degrees Celsius, taken once a minute using a digital thermometer in a laboratory where the air temperature is controlled at a nominal 20 °C. Here there is more variability; there is no obvious pattern over the full half-hour of readings, although for briefer periods some patterns can be seen, such as temporary steady increases in temperature.

5.2.3 Estimation of intercept and slope using least-squares fitting

We often need to estimate the degree of dependence of one quantity upon another. As examples, we may wish to establish in what way

- the length of a metal rod increases as the temperature of the rod increases;
- the diameter of a crater in the Earth formed by a fast-moving object depends on the kinetic energy of the object;

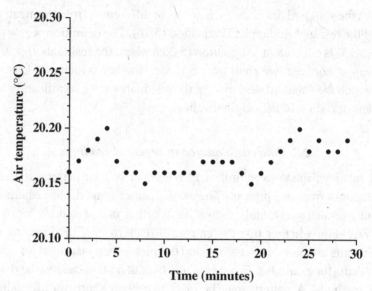

Figure 5.2. Air-temperature readings of a digital thermometer with 0.01-°C resolution.

- the amount of petrol consumed by a car depends on the distance driven; and
- the gravitational force exerted by an object depends on the mass of the object.

The dependence of one quantity upon another may be *linear*; that is, if we plot one quantity against the other on a graph, the plotted points follow a straight line.[10] We shall consider here only linear dependences.

There will be an amount of random scatter or noise superimposed on any dependence between two variables. Figure 4.1 shows the considerable scatter observed in a particular case where the temperature coefficient of a standard resistor was being measured. In other cases the scatter may be so large that it effectively obscures any dependence of one variable on another.

When quantities like resistance and temperature are paired, it is common to refer to one quantity as the *explanatory* or *predictor* variable. For example, the change in temperature 'explains' the change in resistance, and any particular value of temperature (within the range of measurements) 'predicts' the likely value of resistance at that temperature. The other variable is the *response* variable. When graphs are drawn, the explanatory or predictor variable is normally plotted along the x-axis, and the response variable along the y-axis. We note that the presence of one explanatory variable by no means precludes other explanatory variables. For

---

[10] Such is always the case if we are considering only small changes in the quantities, and this is just another way of saying that a section of any curve is approximately a straight line if the section is short enough.

Table 5.2. *Variation of voltage with time*
for a voltage standard

t (years)	V (μV/V)
0.79	2.2
1.89	2.5
3.17	2.8
4.62	3.2
5.96	3.5

example, the amount of petrol consumed by a car depends not only on the distance driven but also on its weight, age and engine efficiency. We shall consider only single explanatory variables.

We assume that the scatter affects only the response variable and that the explanatory variable can be measured without significant error. This is often, but not always, the case. If the explanatory variable is elapsed time, then, because there are very accurate ways of measuring time, we can usually be confident that the measurement of elapsed time has negligible error. When the explanatory variable is some other quantity, such as temperature, we should take care to minimise the errors in it, in order that the following analysis remain valid.

Consider a situation in which we have five measurements of voltage of a nominal 10-V voltage standard. This is a type of portable artefact often used in the electronics industry, and is calibrated against references traceable to the primary national standard of voltage.[11] Ideally a voltage standard should be perfectly stable, but in practice a slow drift and a random scatter are always observed. The five values span an approximately five-year period, and are shown in table 5.2, with time as the explanatory variable and voltage as the response variable. Time, t_i ($i = 1, 2, \ldots, 5$), is given in years, starting from 1 January 1998 as year 0. The voltage, V_i ($i = 1, 2, \ldots, 5$), is given in parts per million above 10 V, or equivalently in μV/V above 10 V. Figure 5.3 shows the plot of voltage against time.

There is a positive drift with time, and we estimate it by fitting a straight line to the points using the least-squares condition. The slope of the line will equal the drift. We write the equation describing the straight line as

$$V = V_0 + bt, \tag{5.35}$$

[11] This is based on the Josephson effect in superconductors (Rose-Innes and Rhoderick 1977). A brief description is given in section 4.1.3.

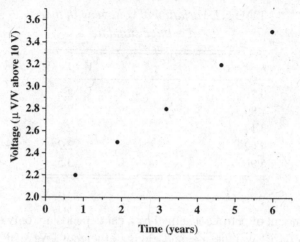

Figure 5.3. Dependence of voltage on time.

where V_0 and b are the two parameters that specify a straight line on a plane. Equation (5.35) may be recognised as having the commonly written form $y = a + bx$ describing a straight line, where a is the intercept on the y-axis and b is the slope. Here we have $a = V_0$. The intercept of the line is V_0 on the vertical (V) axis (when drawn so as to intersect the horizontal (t) axis at $t = 0$) and b is the slope, which is the drift in μV/V per year (μV/V (yr)$^{-1}$). The least-squares condition allows both V_0 and b to be estimated, although in this case b is the parameter of greater interest.

Using a similar procedure to that in section 5.2.1, and following equation (5.35), we write the values in table 5.2 as

$$2.2 = V_0 + 0.79b + \epsilon_1,$$
$$2.5 = V_0 + 1.89b + \epsilon_2,$$
$$2.8 = V_0 + 3.17b + \epsilon_3, \qquad (5.36)$$
$$3.2 = V_0 + 4.62b + \epsilon_4,$$
$$3.5 = V_0 + 5.96b + \epsilon_5,$$

which are five equations with seven unknowns, V_0, b, and $\epsilon_1, \ldots, \epsilon_5$. The least-squares condition enables a unique solution for V_0 and b to be found. The sum of squares, Q, of the five residuals is

$$Q = \epsilon_1^2 + \epsilon_2^2 + \epsilon_3^2 + \epsilon_4^2 + \epsilon_5^2$$
$$= (2.2 - V_0 - 0.79b)^2 + (2.5 - V_0 - 1.89b)^2 + (2.8 - V_0 - 3.17b)^2$$
$$+ (3.2 - V_0 - 4.62b)^2 + (3.5 - V_0 - 5.96b)^2. \qquad (5.37)$$

The required values of V_0 and b are such as to make Q a minimum. To accomplish this the partial derivatives of Q with respect to V_0 and with respect to b are both to be set equal to zero. We have, therefore,

$$\frac{\partial Q}{\partial V_0} = -(2.2 - V_0 - 0.79b) - (2.5 - V_0 - 1.89b) - (2.8 - V_0 - 3.17b)$$
$$- (3.2 - V_0 - 4.62b) - (3.5 - V_0 - 5.96b) \qquad (5.38)$$
$$= 0$$

or

$$5V_0 + b(0.79 + 1.89 + 3.17 + 4.62 + 5.96) = 2.2 + 2.5 + 2.8 + 3.2 + 3.5, \qquad (5.39)$$

leading to

$$5V_0 + 16.43b = 14.2, \qquad (5.40)$$

our first equation.

Also,

$$\frac{\partial Q}{\partial b} = -0.79(2.2 - V_0 - 0.79b) - 1.89(2.5 - V_0 - 1.89b)$$
$$- 3.17(2.8 - V_0 - 3.17b) - 4.62(3.2 - V_0 - 4.62b)$$
$$- 5.96(3.5 - V_0 - 5.96b)$$
$$= 0, \qquad (5.41)$$

which simplifies to

$$16.43V_0 + 71.111b = 50.983. \qquad (5.42)$$

This is the second equation for V_0 and b. Equations (5.40) and (5.42) are two equations with two unknowns, V_0 and b, and have solutions

$$V_0 = 2.010\,58\,\mu V/V, \qquad b = 0.252\,41\,\mu V/V\,(yr)^{-1}. \qquad (5.43)$$

These values would normally be reported to fewer decimal places, for example $b = 0.25\,\mu V/V(yr)^{-1}$, but to avoid round-off error it is worth keeping the extra decimal places for the calculations that follow. Figure 5.4 is a repeat of figure 5.3, but with the line of best fit included. The slope of the line, $0.252\,41\,\mu V/V\,(yr)^{-1}$, indicates a positive drift of about one part per million in four years, and a quick inspection of the original measured values shows this to be a plausible result.

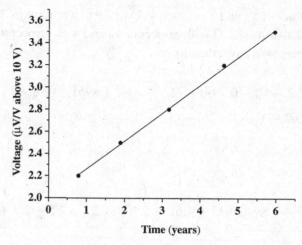

Figure 5.4. A straight line fitted to the data in table 5.2.

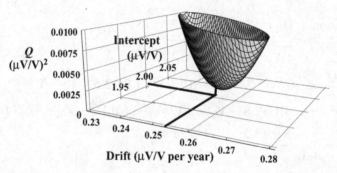

Figure 5.5. The sum of squares, Q, as a function of drift, b, and intercept, V_0.

The second derivatives $\partial^2 Q/\partial V_0^2$ and $\partial^2 Q/\partial b^2$ obtained from equations (5.38) and (5.41) are both found to be positive, confirming that the values obtained for V_0 and b make Q a minimum.[12]

Figure 5.5 is a three-dimensional counterpart to figure 5.1. In figure 5.5, the sum of squares, Q, is displayed as a function both of drift, b, and of the intercept, V_0. The shape is a paraboloid, whose bottom is at a height 0.001 133 above the (b, V_0) plane. This minimum height can be determined by substituting the values for b and V_0 shown in equation (5.43) into equation (5.37).

---

[12] An *extremum* value (maximum or minimum) exists if $(\partial^2 Q/\partial V_0^2)(\partial^2 Q/\partial b^2) > [\partial^2 Q/(\partial V_0\,\partial b)]^2$. It may be checked that $\partial^2 Q/\partial V_0^2 = 5$, $\partial^2 Q/\partial b^2 \sim 71.11$ and $\partial^2 Q/(\partial V_0\,\partial b) \sim 16.43$, so the inequality is satisfied. A positive value of $\partial^2 Q/\partial V_0^2$ therefore implies a positive value of $\partial^2 Q/\partial b^2$, and both indicate a minimum.

By substituting the values in equation (5.43) into equation (5.36), we can determine that the residuals have the following values, in $\mu V/V$:

$$\epsilon_1 = -0.009\,98,$$
$$\epsilon_2 = +0.012\,36,$$
$$\epsilon_3 = -0.010\,72, \tag{5.44}$$
$$\epsilon_4 = +0.023\,28,$$
$$\epsilon_5 = -0.014\,94.$$

Again, these sum to zero:

$$\epsilon_1 + \epsilon_2 + \epsilon_3 + \epsilon_4 + \epsilon_5 = 0 \tag{5.45}$$

and it may be checked that this is guaranteed by equation (5.38). However, the residuals are also linked by a second constraint, which follows from equation (5.41). In fact, it is easily checked that

$$0.79\epsilon_1 + 1.89\epsilon_2 + 3.17\epsilon_3 + 4.62\epsilon_4 + 5.96\epsilon_5 = 0. \tag{5.46}$$

Since the five residuals are now constrained by equations (5.45) and (5.46), the residuals have $5 - 2 = 3$ degrees of freedom. Whenever a straight line is fitted by least-squares, the residuals have two degrees of freedom fewer than the number of original values, so in contrast to equation (5.21) we now have

$$\nu = n - 2. \tag{5.47}$$

Using the values of the residuals in equation (5.44), and also equation (5.23), the unbiased estimate of the variance, s^2, of the population from which this sample of residuals is drawn is (in $(\mu V/V)^2$)

$$s^2 = \frac{(-0.009\,98)^2 + (+0.012\,36)^2 + (-0.010\,72)^2 + (+0.023\,28)^2 + (-0.014\,94)^2}{3}.$$
$$\tag{5.48}$$

Thus

$$s^2 = \frac{0.001\,132}{3} = 0.000\,377\,5. \tag{5.49}$$

The standard deviation, s, is therefore $\sqrt{0.000\,377\,5}\ \mu V/V = 0.019\ \mu V/V$. An alternative name for s is the 'root-mean-square' or rms residual, or 'rms scatter'. The value of s is a measure of the 'closeness of fit', since the more accurately the measured points (in figure 5.3) follow a straight line, the smaller are the residuals about that line.[13]

Exercise B
Show that the standard deviation of the voltage values in table 5.2 is 0.522 $\mu V/V$.

The above results lead us to the following general equations for fitting a straight line. Suppose that there are n measured points $(x_i, \ y_i)$ $(i = 1, 2, \ldots, n)$. In the above example, the x_i were values of time and the y_i were values of voltage. The straight line to be fitted may be described as

$$y = a + bx, \tag{5.50}$$

where a is the intercept of the line on the y-axis and b is the slope of the line. It is straightforward to check (using, for example, the numerical values above) that a and b are given by the following formulas. We first define the quantity D, as follows:

$$D = n \sum_{i=1}^{n} x_i^2 - \left(\sum_{i=1}^{n} x_i \right)^2. \tag{5.51}$$

Then

$$a = \frac{\sum_{i=1}^{n} y_i \sum_{i=1}^{n} x_i^2 - \sum_{i=1}^{n} x_i y_i \sum_{i=1}^{n} x_i}{D} \tag{5.52}$$

and

$$b = \frac{n \sum_{i=1}^{n} x_i y_i - \sum_{i=1}^{n} x_i \sum_{i=1}^{n} y_i}{D}. \tag{5.53}$$

The residuals, ϵ_i $(i = 1, 2, \ldots, n)$, are calculated as

$$\epsilon_i = y_i - a - bx_i \qquad (i = 1, 2, \ldots, n) \tag{5.54}$$

and their root-mean-square value as

$$s = \sqrt{\frac{\sum_{i=1}^{n} \epsilon_i^2}{n-2}}. \tag{5.55}$$

---

[13] We note that this scatter about the line of best fit is much less than the standard deviation of the original voltages in table 5.2. This is in sharp contrast to the previous example where only the mean was estimated. In that example the standard deviation of the residuals, or the rms scatter, was identical to the standard deviation of the original values.

Table 5.3. *Concentration-versus-area data for analysis of sodium chloride by HPLC*

Concentration (x) (ppm)	Area (y) (arbitrary units)
1.028	1.59×10^4
2.056	3.10×10^4
5.141	6.34×10^4
7.711	9.91×10^4
10.282	1.27×10^5
15.422	2.01×10^5
25.704	3.83×10^5

Example 1

High-performance liquid chromatography (HPLC) is used to establish the concentration of an analyte, such as sodium chloride, in solution. To accomplish this, an HPLC instrument is first calibrated using known concentrations of the analyte. Table 5.3 shows the response of an instrument (which is the area under an absorption peak detected by the instrument) for various concentrations (in parts per million, 'ppm') of sodium chloride. Assuming that equation (5.50) applies to these data, use least-squares to determine the intercept, a, and the slope, b.

Answer

In order to calculate the intercept and slope we need to determine the sums in equations (5.51)–(5.53), i.e.

$$\sum_{i=1}^{n} x_i = 67.344, \qquad \sum_{i=1}^{n} y_i = 920\,400, \qquad \sum_{i=1}^{n} x_i y_i = 15\,420\,448.7,$$

$$\sum_{i=1}^{n} x_i^2 = 1095.426\,546.$$

Substituting these sums into equation (5.51) (and noting that $n = 7$) gives

$$D = 7 \times 1095.426\,546 - (67.344)^2 = 3132.771\,486.$$

Now, using equations (5.52) and (5.53), we have

$$a = \frac{920\,400 \times 1095.426\,546 - 15\,420\,448.7 \times 67.344}{3132.771\,486} = -9654.1,$$

$$b = \frac{7 \times 15\,420\,448.7 - 67.344 \times 920\,400}{3132.771\,486} = 14\,670.6 \text{ ppm}^{-1}.$$

Table 5.4. *Variation of the acceleration*
due to gravity with height

Height (x) (km)	Acceleration (y) (m/s^2)
10.0	9.76
20.0	9.73
30.0	9.70
40.0	9.68
50.0	9.64
60.0	9.58

The estimates of intercept and slope calculated in this example are given to an excessive number of significant figures, but until the standard uncertainty in each has been determined, it is not possible to decide how many figures should be displayed.

Exercise C

The acceleration due to gravity, g, near the Earth's surface depends on several factors including the height above the Earth's surface at which the measurement is made. Table 5.4 contains values of g obtained at several heights above the Earth's surface.

(1) Assuming that equation (5.50) is valid for the data in table 5.4, determine

$$\text{(i)} \sum_{i=1}^{n} x_i, \qquad \text{(ii)} \sum_{i=1}^{n} y_i, \qquad \text{(iii)} \sum_{i=1}^{n} x_i y_i, \qquad \text{(iv)} \sum_{i=1}^{n} x_i^2.$$

(2) Use the summations in part (1) to determine the best estimate of the intercept and slope of a straight line through the (x, y) data.

Calculations of the summations in equations (5.51)–(5.53) are most efficiently accomplished using a computer-based spreadsheet, such as Excel by Microsoft. This spreadsheet has built-in functions that allow direct fitting by least-squares. Many scientific calculators possess equivalent built-in functions.

5.2.4 Standard uncertainties of estimates

Using the rms value, s, of the residuals as in equations (5.23) and (5.55), we can calculate the standard uncertainties of the estimates themselves. The standard uncertainty, $s_{\bar{x}}$, of the *mean*, for n mutually uncorrelated values is given by[14]

$$s_{\bar{x}} = \frac{s}{\sqrt{n}}. \qquad (5.56)$$

$s_{\bar{x}}$ is, therefore, less than s by a factor $\sqrt{n}$. In the example in section 5.2.1 of the six pieces of fruit, where $s = 0.18$ g, we therefore have $s_{\bar{x}} = 0.18$ g$/\sqrt{6} = 0.073$ g.

Whereas the standard deviation or standard uncertainty of the original values, s, changes little whether we take few or many measurements, the standard uncertainty in the mean, $s_{\bar{x}}$, *decreases* with the number of (uncorrelated) measurements. This is the statistical underpinning of our intuitive notion (which is not always correct!) that the more measurements we take, the more accurate the result. We note the square-root dependence; thus taking 50 measurements instead of 5 should reduce the standard uncertainty of the mean by a factor of only $\sqrt{10} \simeq 3$. Notably, if the dominant error in our measurements is a systematic error, there will be little or no benefit to be gained by taking many measurements.

In using equation (5.56), an important proviso is that the n residuals should be uncorrelated. This will be satisfied if the values are independent. As a test of independence, the values should be examined to check whether they follow a pattern, for example a drift or oscillation. If they do, they are not independent values and the effective n in equation (5.56) may be less than the number of measured values. If there is a perfectly steady drift, the effective n in equation (5.56) is 1, and in such a case[15] it would be more appropriate to use $n = 1$ in equation (5.56), or alternatively to fit a straight line, as described in section 5.2.3. The case of correlated readings will be discussed further in section 7.2.

When a straight line is fitted to data, the standard uncertainties of the intercept, a, and slope, b, are[16]

$$s_a = s\sqrt{\frac{\sum_{i=1}^{n} x_i^2}{D}} \qquad (5.57)$$

and

$$s_b = s\sqrt{\frac{n}{D}}, \qquad (5.58)$$

where s and D are given in equations (5.55) and (5.51), respectively.

---

[14] This was discussed in section 4.3.
[15] This also applies to cases where there is some scatter about an obvious drift.
[16] See, for example, Bevington and Robinson (2002).

Example 2

Using the data in table 5.3, show that the standard uncertainties in the intercept and slope are given by $s_a = 8070$ and $s_b = 645$ ppm^{-1}.

Answer

From the solution to example 1, we have

$$\sum_{i=1}^{n} x_i^2 = 1095.426\,546, \qquad D = 3132.771\,486.$$

In order to determine s, we use equation (5.55) with $n = 7$ and ϵ_i given by

$$\epsilon_i = y_i - (-9654.1) - (14\,670.6)x_i \qquad (i = 1, 2, \ldots, n).$$

This gives $s = 13\,646.7$. Now substituting $\sum_{i=1}^{n} x_i^2 = 1095.426\,546$, $D = 3132.771\,486$ and $s = 13\,646.7$ into equations (5.57) and (5.58) gives

$$s_a = 13\,646.7\sqrt{\frac{1095.426\,546}{3132.771\,486}} = 8070$$

and

$$s_b = 13\,646.7\sqrt{\frac{7}{3132.771\,486}} = 645.$$

Exercise D

Using the data in table 5.4, calculate the standard uncertainties in the intercept and slope of the best-fit line through the data.

Returning to the data in table 5.2, we can use equations (5.57) and (5.58) to show that the standard uncertainties in the the the intercept, s_{V_0}, and drift, s_b, are given, respectively, by $s_{V_0} = 0.017\,71$ μV/V and $s_b = 0.004\,70$ μV/V (yr)$^{-1}$.

The standard uncertainty, s_b, of the drift (that is, the slope) is much smaller (in absolute magnitude) than the drift, b, itself. In fact, the ratio b/s_b is 0.252/0.004 70 or about 54. We can therefore conclude that (as figure 5.4 indicates) there is a very easily observable drift, or, expressing this in another way, the random scatter in the measurement, although it evidently exists, is much too small to obscure the drift. In statistical language we say that the drift is highly *significant*.

In the example of the temperature variation of the resistance of a standard resistor, shown in figure 5.6, a much larger scatter about the line of best fit is observed. Here we have $s = 0.59$ μΩ/Ω, $b = 0.071$ μΩ/Ω (°C)$^{-1}$ and $s_b = 0.037$ μΩ/Ω (°C)$^{-1}$. The temperature coefficient of resistance is only about twice its standard uncertainty. Although the temperature coefficient is significant, this significance evidently is

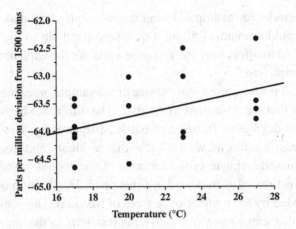

Figure 5.6. The variation with temperature of the resistance of a standard resistor.

more provisional than in the case of the measurement of the drift of the voltage standard.

Like $s_{\bar{x}}$, both s_{V_0} and s_b tend to decrease as the square root of n. Again, the proviso is that the residuals should be uncorrelated, and they will be uncorrelated if they are independent. Any pattern among the residuals will imply a lack of independence, and equations (5.57) and (5.58) will not hold. We might then consider fitting a higher-order curve, say a quadratic parabola $y = a + bx + cx^2$, to the original values. The coefficients, a, b and c, can be determined by least-squares using a similar procedure to that described in section 5.2.3 for a and b; as might be expected, the relation between the number of degrees of freedom ν and number of points, n, is now $\nu = n - 3$. There are also cases where a polynomial is not appropriate, and where we should try to fit an exponential relationship. Thus, if a variable decays in time with a 'time-constant' τ (a frequent case in electronics, where the variable may be the voltage across a capacitor) or with a 'half-life' t_h (referring to a radioactive isotope), the response variable in question varies as $y = y_0 e^{-t/\tau}$ or $y = y_0 e^{-(t \log 2)/t_h}$. The technique of least-squares can be adapted to suit these and similar cases.[17]

5.2.5 Further remarks on least-squares fitting

The least-squares approach allows us to extract estimates of one or more parameters from the data (thus, for linearly related data, we can find best estimates of the slope and intercept of a line through the data). More complicated cases of least-squares

---

[17] For a comprehensive guide to fitting by least-squares, refer to Kutner *et al.* (2004).

fitting would include, for example, fitting an intercept, slope and rate of change of slope; this would amount to fitting a quadratic parabola to the data. Random errors are assumed to affect only the response variable; the explanatory variable is assumed to be error-free.[18]

In every case, it is understood that the size of the sample, n, exceeds the number, q, of parameters that we wish to fit to the data. The difference between n and q is the number, v, of degrees of freedom of our least-squares fit: $v = n - q$. Having performed the least-squares fit, we are left with n residuals, and from these we can calculate an unbiased estimate of the variance of the population of the residuals. This variance is estimated as indicated in equation (5.23): the sum of squares of all n residuals, divided by the number of degrees of freedom. The standard deviation of the fit, or 'root-mean-square' residual (rms residual), is the square root of this variance. When $q = 1$, the single estimate is the mean, and the standard deviation of this fit is none other than the ordinary standard deviation as defined in equation (5.13), with $v = n - q = n - 1$ in the denominator.

The smaller the number of degrees of freedom, the less reliable our least-squares fit. Imagine an extreme case where the sample size is only two ($n = 2$), and we wish to fit an intercept and drift ($q = 2$) to these two values. In this case the number of degrees of freedom is zero ($v = n - q = 0$), implying a totally unreliable fit. This makes sense: we cannot hope to fit a straight line with credible slope and intercept to just two points. In fact, with two points it is always possible to draw a straight line through both of them exactly, giving what might naively be imagined to be a perfect fit. However, there is no 'redundancy' here.[19] For a reliable fit, more than two points are required, and the more points the better, giving more degrees of freedom. We need, in other words, more 'redundancy'; the greater the number of points, the better our protection against inevitable random errors (and we are also better able to assess their influence on our results). If the size of the sample is no greater than the number of parameters we wish to fit to it, we are completely exposed to the effect of random errors, and the fit will be useless. Equation (5.23) expresses this unfortunate situation as the indeterminate quantity zero divided by zero; the numerator in equation (5.23) is zero since all the ϵ_i are zero (the fit being 'perfect'), but so is the denominator,[20] v.

---

[18] A more complicated procedure, sometimes called total least-squares, may be used when the explanatory variables also have random errors. For more information about fitting in such cases, see Macdonald and Thompson (1992) and Balsamo *et al.* (2005).

[19] In statistics and metrology 'redundant' does not mean 'useless' or 'unnecessary', but rather something akin to 'generous'.

[20] However, there are exceptional cases. If a quantity is measured only twice, at the beginning and end of an interval, and if we know from prior evidence that the uncertainty in each measurement is much less than the magnitude of the change in the quantity, then we may have confidence in the measured amount of change.

Two types of estimation have been involved in the least-squares fitting to a sample of n values of data. We first estimate the parameters, by minimising the sum of squares of the residuals. After the parameters have been determined, and therefore also the residuals, we can calculate the unbiased estimate, s^2, of the variance of the population of the residuals, using equation (5.23) with $v = n - q$. This is the second type of estimation. In fact both types of estimation provide unbiased estimates; for example, $E(\bar{x}) = \mu$ as stated in equation (5.2), and it can also be shown that the expectation values of intercept and slope are the values of intercept and slope for the population from which the sample was drawn.

From the unbiased estimate, s^2, of variance and its square root, we can estimate the standard uncertainty of the estimated parameters themselves. This standard uncertainty is of the order of $\sqrt{n}$ less than s, as expressed by, for example, equations (5.56)–(5.58), provided that the residuals are uncorrelated. (For a single estimated parameter, the mean, this proviso is equivalent to the original sampled values being uncorrelated).

5.3 Covariance and correlation

Suppose that there is a significant linear dependence of y on x, so that in the fitted equation of a straight line, $y = a + bx$, the value of b is significant (meaning that b is considerably greater in absolute magnitude than its own standard uncertainty). Then, as might be expected, x and y have a significant mutual correlation. The *linear correlation coefficient*, r, is defined as follows. If there are n pairs x_i, y_i ($i = 1, 2, \ldots, n$), we first define the *covariance* of x and y, as estimated for the populations of x and of y, as

$$\text{covariance}(x, y) = \frac{\sum_{i=1}^{n}(x_i - \bar{x})(y_i - \bar{y})}{n - 1}, \tag{5.59}$$

where $\bar{x}$ and $\bar{y}$ are the means of the x's and y's, respectively. We now express r as follows:

$$r = \frac{\text{covariance}(x, y)}{\sqrt{\text{variance of } x \times \text{variance of } y}} \tag{5.60}$$

or, more simply,

$$r = \frac{\text{covariance}(x, y)}{\text{standard deviation of } x \times \text{standard deviation of } y}, \tag{5.61}$$

where the variances and standard deviations are taken as estimated over the population.[21] The variance of x is given by

$$\text{variance of } x = \frac{\sum_{i=1}^{n}(x_i - \bar{x})^2}{n - 1}$$

and similarly for the variance of y.

We can therefore define r as

$$r = \frac{\sum_{i=1}^{n}[(x_i - \bar{x})(y_i - \bar{y})]}{\sqrt{[\sum_{i=1}^{n}(x_i - \bar{x})^2][\sum_{i=1}^{n}(y_i - \bar{y})^2]}}. \tag{5.62}$$

The same equation would be obtained if the covariance were defined with the divisor, n, and similarly the standard deviations. Whether n or $n - 1$ is chosen is immaterial when calculating the correlation coefficient. Equation (5.62) also implies that the correlation between x and y is identical to that between y and x.

r is a dimensionless quantity, since equation (5.62) shows that its dimensions are $x \times y$ divided by $\sqrt{x^2 \times y^2}$. It may be shown that r must lie between -1 (perfect negative correlation) and $+1$ (perfect positive correlation).[22] A positive slope of the line of best fit implies a positive correlation, r, and conversely a negative slope implies a negative r. The greater the scatter around the line of best fit, the closer r will approach zero. If this scatter is zero, r will then equal $+1$ or -1, depending, respectively, on whether the slope is positive or negative, but independently of the actual value of the slope (unless the slope is exactly zero; for zero slope and scatter, r is indeterminate).

There is a distinction between independence and zero correlation. It is possible for two variables, x and y, to have zero mutual correlation, yet to be mutually dependent. For example, if x and y are related by the equation $x^2 + y^2 = 1$, so that x and y lie on the circumference of a circle of radius 1, it may be shown that the correlation between x and y is zero. (Thus the four points with x, y coordinates $(1, 0)$, $(0, 1)$, $(-1, 0)$ and $(0, -1)$ lie on this circle, and their mutual correlation $r = 0$.) However, x and y are not mutually independent, since they are related by the equation $x^2 + y^2 = 1$. In fact, independence implies zero correlation, but zero correlation does not imply independence.

Equation (5.6) gives the population variance expressed as an expectation function. If we now take μ_x and μ_y as the means of the populations of the x's and y's, respectively, the covariance between the populations can be written, analogously to equation (5.6), as

$$\text{covariance}(x, y) = E[(x_i - \mu_x)(y_i - \mu_y)]. \tag{5.63}$$

---

[21] See, for example, equation (5.8).
[22] See, for example, chapter 3 in Wilks (1962).

Comparing equation (5.63) with equation (5.6) shows that the covariance of a quantity with itself is simply the variance of that quantity.

Expanding the right-hand side of equation (5.63) gives

$$\text{covariance}(x, y) = E(x_i y_i) - \mu_y E(x_i) - \mu_x E(y_i) + \mu_x \mu_y$$
$$= E(x_i y_i) - \mu_x \mu_y \tag{5.64}$$

since $E(x_i) = \mu_x$ and $E(y_i) = \mu_y$. Equation (5.64) shows that a covariance may be regarded as the expectation of a product, minus the product of the expectations. This is analogous to the interpretation of a variance expressed by equation (5.11), namely the mean square minus the squared mean.

If the two populations are uncorrelated, then[23] equation (5.63) factorises into $E(x_i - \mu_x)E(y_i - \mu_y)$ and each factor is zero (see equation (5.3)). Uncorrelated populations have zero covariance, and therefore (of course!) zero correlation.

In terms of expectation functions, r may be written as

$$r = \frac{E[(x_i - \mu_x)(y_i - \mu_y)]}{\sqrt{E[(x_i - \mu_x)^2]E[(y_i - \mu_y)^2]}}. \tag{5.65}$$

Using equations (5.6) and (5.63), equation (5.65) may be written as

$$r = \frac{E[(x_i - \mu_x)(y_i - \mu_y)]}{\sigma_x \sigma_y} = \frac{\text{covariance}(x, y)}{\sigma_x \sigma_y}, \tag{5.66}$$

so the correlation coefficient, a dimensionless quantity, may be regarded as a 'normalised covariance', the term 'normalised' used here as implying that a quantity has been scaled appropriately so as to be dimensionless. Here the scaling factor is the product of the standard deviations of the two populations.

We have noted above that the covariance of a quantity with itself is the variance of that quantity. Equation (5.65), in which a covariance is divided by the square root of a product of the two variances, indicates that the correlation coefficient of a quantity with itself is $+1$.

5.3.1 Correlation between two linearly related variables, without random error

Equation (5.62) can be used to illustrate the correlation between variables x and y when they are linearly related by $y = a + bx$, without any random error. Then, for our sample of size n, it follows that

$$\bar{y} = a + b\bar{x}, \tag{5.67}$$

---

[23] Using property (d) of the expectation function in section 5.1.1.

where $\bar{x}$ and $\bar{y}$ are, respectively, the means of the x and y values. Then

$$y_i - \bar{y} = (a + bx_i) - (a + b\bar{x}) = b(x_i - \bar{x}). \tag{5.68}$$

Equation (5.68) implies that

$$\sum_{i=1}^{n}(x_i - \bar{x})(y_i - \bar{y}) = b\sum_{i=1}^{n}(x_i - \bar{x})^2, \tag{5.69}$$

and equation (5.62) therefore reads

$$r = \frac{b\sum_{i=1}^{n}(x_i - \bar{x})^2}{\sqrt{\left[\sum_{i=1}^{n}(x_i - \bar{x})^2\right]\left[b^2\sum_{i=1}^{n}(x_i - \bar{x})^2\right]}} = \pm 1, \tag{5.70}$$

the sign being positive if the slope b is positive, and negative if b is negative.

Unless the slope is zero, therefore, x and y are perfectly mutually correlated (whether positively or negatively). (For zero slope, r is not defined.)

5.3.2 Correlation between two linearly related variables, with random error

In section 5.3.1, we neglected random error. We now suppose that the relationship between x and y is, more realistically, given by

$$y_i = a + bx_i + \epsilon_i, \tag{5.71}$$

where ϵ is the random error, which is assumed to be uncorrelated with x and with y, and (without loss of generality) taken to have zero mean. Then $\bar{y} = a + b\bar{x}$ as before, but now we have

$$\begin{aligned} y_i - \bar{y} &= (a + bx_i + \epsilon_i) - (a + b\bar{x}) \\ &= b(x_i - \bar{x}) + \epsilon_i, \end{aligned} \tag{5.72}$$

which is to be compared with equation (5.68).

Equation (5.72) gives

$$\begin{aligned} \sum_{i=1}^{n}(y_i - \bar{y})^2 &= \sum_{i=1}^{n}[b(x_i - \bar{x}) + \epsilon_i]^2 \\ &= \sum_{i=1}^{n}b^2(x_i - \bar{x})^2 + \sum_{i=1}^{n}\epsilon_i^2 + 2b\sum_{i=1}^{n}\epsilon_i(x_i - \bar{x}). \end{aligned} \tag{5.73}$$

We now assert that, in the third term on the right-hand side of equation (5.73), $\sum_{i=1}^{n}\epsilon_i(x_i - \bar{x})$ is zero or very close to zero. This is reasonable if the ϵ_i are uncorrelated with the x_i (and hence with $x_i - \bar{x}$), as has been assumed. We also have, with the aid of equation (5.72),

$$\sum_{i=1}^{n}[(x_i - \bar{x})(y_i - \bar{y})] = \sum_{i=1}^{n}\{(x_i - \bar{x})[b(x_i - \bar{x}) + \epsilon_i]\}$$

$$= b\sum_{i=1}^{n}(x_i - \bar{x})^2, \tag{5.74}$$

using again the same assumption. Equations (5.73) and (5.74) inserted into equation (5.62) give

$$r = \frac{b\sum_{i=1}^{n}(x_i - \bar{x})^2}{\sqrt{\sum_{i=1}^{n}(x_i - \bar{x})^2\left[\sum_{i=1}^{n}b^2(x_i - \bar{x})^2 + \sum_{i=1}^{n}\epsilon_i^2\right]}}, \tag{5.75}$$

and dividing the numerator and denominator of equation (5.75) by $b\sum_{i=1}^{n}(x_i - \bar{x})^2$ gives

$$r = \frac{1}{\sqrt{1 + \dfrac{\sum_{i=1}^{n}\epsilon_i^2}{b^2\sum_{i=1}^{n}(x_i - \bar{x})^2}}}. \tag{5.76}$$

The additive constant a, in equation (5.72), does not appear in the expression for r in equation (5.76). This illustrates a general rule: when calculating correlations, the presence of an additive constant has no effect on the correlation.[24]

Equation (5.76) indicates that the greater the random error affecting a linear relationship, the closer will be the approach of r to zero. Additionally, the smaller the slope, b, for a given amount of random error, the closer will be the approach of r to zero.

Example 3
The data in table 5.5 were obtained during the calibration of an atomic absorption spectrometer using standard silver solutions of various concentrations. Assuming that the relationship given by equation (5.50) is valid for the data in table 5.5, calculate

(a) the intercept, a, and the slope, b, of the best line through the data; and
(b) the correlation coefficient, r.

---

[24] Neither does a multiplicative constant. In general, if from prior knowledge the correlation coefficient between x and y is r, then the correlation coefficient between $Ax + K_1$ and $By + K_2$ is also r. Here A, B, K_1 and K_2 are constants (A and B non-zero).

Table 5.5. *Variation of absorbance with concentration*
for silver solutions

Concentration (x) (ng/mL)	Absorbance (y) (arbitrary units)
0.00	0.002
5.03	0.131
10.10	0.255
15.07	0.391
20.11	0.502
25.06	0.622
30.12	0.766

Answer

(a) Upon applying equations (5.51)–(5.53), we obtain $a = 0.003\,431$ and $b = 0.025\,07$ mL/ng.

(b) Using equation (5.76), where $\epsilon_i = y_i - (a + bx_i)$, we obtain $r = +0.999\,68$.

Exercise E

Show for the data in table 5.2 that the correlation between voltage and time is $r = +0.999\,48$.

5.4 Review

In this chapter we have considered several concepts required when dealing with variability in data such as expectation, variance, correlation and covariance. With these concepts we are able to calculate standard uncertainties when the variability in experimental data is due to random errors. In chapter 6 we turn our attention to uncertainties resulting from systematic errors.

6

Systematic errors

A systematic error causes a measured value to be consistently greater or less than the true value. The amount by which the value differs from the true value may be a constant. Such a situation would occur, for example, when using a micrometer that has a 'zero error': the scale of the micrometer indicating a non-zero value when the jaws of the micrometer are closed. In other circumstances, a systematic error may be proportional to the magnitude of the quantity being measured. For example, if a wooden metre rule has expanded along its whole length as a consequence of absorbing moisture, the size of the systematic error is not constant but increases with the size of the object being measured.

Systematic errors may be revealed in two ways: by means of specific information or when the experimental set-up is changed (whether intentionally in order to identify systematic errors, or for some other reason). In both cases we need a good understanding of the science underlying the measurement. In general, statistical analysis may or may not be involved in assessing the uncertainty associated with a systematic error, so this uncertainty may be Type A or B. When the effect of random errors has been minimised, for example by taking the mean of many values, the influence of systematic errors remains unless they too have been identified and corrected for.

Since a systematic error does not necessarily cause measured values to vary, it often remains hidden (and may be larger than the random errors). Experienced experimenters consistently review their methods in an effort to identify and quantify systematic errors.

6.1 Systematic error revealed by specific information

All instruments and artefacts have a systematic error, which may or may not be significant, depending on the particular application. When they are calibrated against a standard (which, by definition, is a more accurate instrument or artefact for that application), the systematic error will be revealed, and the analysis of the calibration will also provide an estimate of the uncertainty to be associated with

Table 6.1. *An extract from the calibration report on an aneroid barometer*

Instrument reading (mbar)	True pressure (mbar)	Instrument correction (mbar)
960.33	960.00	−0.33
970.36	970.00	−0.36
980.37	980.00	−0.37
990.36	990.00	−0.36
1000.38	1000.00	−0.38
1010.37	1010.00	−0.37
1020.35	1020.00	−0.35
1030.40	1030.00	−0.40
1040.28	1040.00	−0.28
1050.37	1050.00	−0.37

The aneroid barometer was compared against a standard pressure balance, over a barometric pressure range from 960 to 1050 mbar, with the results shown in the table. The instrument was not adjusted. The temperature of the test was 20.3 °C (±0.3 °C). When the sign of the correction is positive (+), the correction should be added to the observed reading to give the correct pressure; and, when it is negative (−), subtracted from the reading. The corrections in table 6.1 are given to the nearest 0.01 mbar with an uncertainty of ±0.08 mbar.

that systematic error. After calibration, the systematic error is effectively removed through the procedure of applying the relevant *correction* to the indicated reading. What remains, after this correction has been applied, is the uncertainty associated with the systematic error.

An instrument that displays a more positive reading than it should is conventionally regarded as having a positive systematic error, equal to the difference between the displayed and correct readings. The *correction* that cancels out the systematic error then has a negative sign. A similar convention applies to the value provided by an artefact.

The two central items of information are, therefore, the systematic error and the uncertainty associated with this error. A calibration report on the instrument or artefact issued by an accredited calibrating authority will invariably state both items. The uncertainty is Type B, since the act of reading the report involves no statistical analysis. This is the most straightforward case of a systematic error and its associated Type B uncertainty.

6.1.1 An example of assessing uncertainty using a calibration report

Table 6.1 shows an extract from a calibration report on an aneroid barometer.[1] (A mbar (millibar) is a unit of atmospheric pressure equivalent to 1 hPa (one

---

[1] The report was issued by, and an extract is printed here by courtesy of, the National Measurement Institute of Australia.

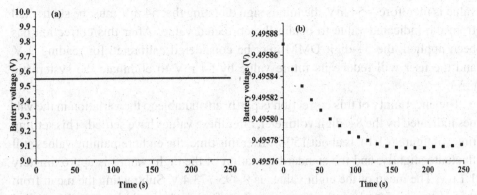

Figure 6.1. (a) Variation of battery voltage with time. (b) Using higher resolution, and revealing a systematic error in the $3\frac{1}{2}$-digit DMM.

hectopascal or 100 pascals)). If the barometer reads, say, 990.1 mbar and we ignore the calibration report, the consequent systematic error will be $+0.36$ mbar (we interpolate, to sufficient accuracy, between the tabulated instrument readings of 980.37 mbar and 990.36 mbar). After we have applied the correction of -0.36 mbar, obtaining 989.74 mbar as the corrected value of the barometer reading, the uncertainty of the value 989.74 mbar is 0.08 mbar, as stated in the report.[2] The $\pm$ symbol indicates that the actual value is likely to be somewhere within the range $(989.74 - 0.08)$ mbar to $(989.74 + 0.08)$ mbar, that is 989.66 mbar to 989.82 mbar. The uncertainty is Type B from our point of view.

Exercise A
The aneroid barometer discussed in this section indicates a pressure of 1035 mbar. Using the data in table 6.1, apply the appropriate correction to this value and state the uncertainty in the corrected value.

6.1.2 The example of correction to values displayed by a digital multimeter (DMM)

The procedure of calibration of an instrument or artefact usually involves both a Type A and a Type B estimation of uncertainty. We illustrate this by considering the calibration of the $3\frac{1}{2}$-digit DMM in figure 6.1(a) set on its 20-V range against the much more accurate $8\frac{1}{2}$-digit DMM in figure 6.1(b). These values are obtained simultaneously, with a battery of nominal emf equal to 9 V connected to both DMMs. This procedure reveals that, when measuring a voltage around 9 V, the $3\frac{1}{2}$-digit DMM reads about 54 mV too high. The correction to be applied to its

---

[2] Strictly, the 0.08-mbar uncertainty in this example should be termed an *expanded uncertainty*. Expanded uncertainty will be considered in detail in chapter 10.

value is therefore -54 mV, the minus sign dictating that 54 mV must be subtracted from the indicated value to obtain the corrected value. After this correction has been applied, the $3\frac{1}{2}$-digit DMM may be considered calibrated for reading 9 V, and the user will reduce its future value by 54 mV to eliminate the systematic error.

The uncertainty of this correction is partly attributable to the variation in the values indicated by the $8\frac{1}{2}$-digit voltmeter, after these values have settled. This settling time, in figure 6.1(b), is about 150 s. After this time, the eight remaining values still fluctuate slightly, and the standard deviation of the eight values is approximately 1.1 μV. The mean of the eight values is 9.495 773 4 V. Subtracting the mean from 9.55 V gives -54 mV as the correction (to sufficient decimal places) to be applied to the $3\frac{1}{2}$-digit DMM, with a standard uncertainty of 1.1 μV.

The question which next arises is whether these eight values can be assumed to be uncorrelated, as discussed in section 5.2.4. If the eight values are scattered independently of one another, we can assign a standard uncertainty of $1.1/\sqrt{8}$ μV $\simeq$ 0.4 μV to the mean, and this 0.4 μV is then the standard uncertainty of the -54-mV correction. Since, however, only eight values are taken and, more importantly, since they appear to exhibit a small positive drift – possibly the initial stage of a slow oscillation – it would be prudent to conclude that their lack of correlation has not been established and that a more conservative (that is, more cautious) estimate of the standard uncertainty of the correction is the standard deviation 1.1 μV of the fluctuations. We therefore have a standard uncertainty of 1.1 μV (about one part in 10^7) associated with the correction -54 mV to be applied to the $3\frac{1}{2}$-digit voltmeter when it is used for measurements around 9 V. This standard uncertainty is Type A, since statistical analysis was used for estimating it.

This Type A standard uncertainty should be combined, by root-sum-squares, with the standard uncertainty that is quoted in the latest available calibration report (not shown) on the $8\frac{1}{2}$-digit DMM when reading 9 V. (This calibration report also states the required corrections to be applied to values indicated by the $8\frac{1}{2}$-digit DMM, and these are assumed to have been applied already to the readings plotted in figure 6.1(b)). The calibration report on the $8\frac{1}{2}$-digit DMM states the standard uncertainty as 1 μV, a Type B uncertainty from our point of view, and combining it with the Type A standard uncertainty of 1.1 μV gives a combined standard uncertainty of $\sqrt{1.1^2 + 1^2}$ μV $\simeq$ 1.4 μV associated with the -54-mV correction to the values of the $3\frac{1}{2}$-digit DMM. The 1.4 μV has therefore a Type A (1.1 μV) and a Type B component (1 μV). Since the value of 1.4 μV is now a reported value to be used subsequently as the standard uncertainty in the correction to the $3\frac{1}{2}$-digit DMM, we classify it as Type B. We note that 1.4 μV is a negligible uncertainty to the corrected values obtained using the $3\frac{1}{2}$-digit DMM, and naturally so, since it has been calibrated against a much more accurate DMM.

Table 6.2. *Comparison of values displayed by a $3\frac{1}{2}$-digit DMM and a $5\frac{1}{2}$-digit DMM*

Value displayed by $3\frac{1}{2}$-digit DMM (V)	Value displayed by $5\frac{1}{2}$-digit DMM (V)
1.502	1.497 83
1.502	1.497 88
1.502	1.497 69
1.502	1.497 77
1.502	1.497 81
1.502	1.497 68

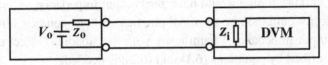

Figure 6.2. The loading effect of a DMM.

Exercise B

A $3\frac{1}{2}$-digit DMM is to be calibrated by comparison with a $5\frac{1}{2}$-digit DMM. Both DMMs are connected simultaneously to a stable voltage source. Table 6.2 shows the values obtained using both DMMs.

(a) Using the data in table 6.2, determine the best estimate of the correction that must be applied to the voltage displayed by the $3\frac{1}{2}$-digit DMM.
(b) Assuming that the values displayed by the $5\frac{1}{2}$-digit DMM are mutually independent, calculate the standard uncertainty of the mean of the values displayed by this DMM.
(c) Given that the calibration report on the $5\frac{1}{2}$-digit DMM states that the standard uncertainty in voltage is 15 μV, calculate the combined standard uncertainty in the correction estimated in part (a).

In the following examples, there is a more obvious need for a good understanding of the scientific background when identifying possible systematic errors.

6.1.3 An example of systematic error due to loading

The voltage output, V_o, of a voltage source with output impedance, Z_o, is measured using a DMM with input impedance Z_i, as indicated in figure 6.2. The DMM terminals are connected to the voltage source.[3] The voltage, V_d, displayed by the

---

[3] The symbol for the voltage source represents a constant ('direct-current' or 'dc') voltage, but the systematic error described here applies also to a varying ('alternating-current' or 'ac') voltage.

DMM is given by

$$V_d = V_o \frac{Z_i}{Z_i + Z_o}. \tag{6.1}$$

Very commonly Z_i is much greater than Z_o (by many orders of magnitude), so that equation (6.1) may be approximated as

$$V_d = V_o \left[1 - \left(\frac{Z_o}{Z_i} \right) \right]. \tag{6.2}$$

V_d is, therefore, less than V_o. Our measurement of V_o has a systematic error $-V_o(Z_o/Z_i)$; and the correction to be applied to the DMM reading is $+V_o(Z_o/Z_i)$.

An 'ideal' voltage source would have zero output impedance, $Z_o = 0$, and this systematic error would then be zero. All practical voltage sources, however, have a non-zero output impedance.[4] Familiarity with the 'loading' effect of a voltage source, as described by equations (6.1) and (6.2), is needed.

The value of Z_i is normally stated in the DMM manufacturer's specifications. We also need to know the value of Z_o of the voltage source, and this is also normally stated in its manufacturer's specifications. It may also happen that Z_o is not given but must be measured separately. The values of both Z_o and Z_i form our specific information for estimating the systematic error of the measurement of V_o. For high-quality DMMs $Z_i \sim 10^{10}\,\Omega$ and $Z_o \sim 1000\,\Omega$ when $V_o \sim 1\,V$, so that the correction in this case is about $+1$ part in 10^7 ($0.1\,\mu V/V$). For high-accuracy measurements at the 1-V level, therefore, we must add $0.1\,\mu V/V$ to the DMM's indicated reading. The uncertainties in Z_o and Z_i determine the uncertainty remaining after this correction of $+0.1\,\mu V/V$ is applied, and this uncertainty will be generally estimated without benefit of statistical analysis as a Type B uncertainty.

A further source of systematic error in this electrical example was stated in section 4.1.3: the zero-offset voltage of a DMM, and the small thermal voltages caused by the Seebeck effect. This will be discussed further in section 6.2.

Exercise C

A $3\frac{1}{2}$-digit DMM is used to measure the output of a voltage source that has an output impedance of $Z_o = 100\,k\Omega$. Assume that the input impedance Z_i of the DMM is $Z_i = 10\,M\Omega$. If the value indicate by the DMM is 1.544 V, what correction must be applied to this value to account for the DMM's finite internal impedance?

---

[4] There exist superconducting voltage sources called Josephson junctions, which have zero output resistance.

Figure 6.3. Nineteen standard weights, ranging from 1 kg to 1 mg (courtesy of the National Measurement Institute of Australia).

6.1.4 Systematic error in weighing due to buoyancy

There are cases where, for the sake of simplicity and convenience, no correction is made for a known small systematic error. On the contrary, the systematic error is tolerated (except when very high accuracy is required), and the quantity that contains it may be given a special name to distinguish it from the corresponding error-free quantity.

An example of this approach occurs when objects are weighed. The relevant standards are mass standards or 'standard weights', shown in figure 6.3, and the mass of an object is determined by comparing its weight with a standard weight using a balance or scales. High-accuracy mass standards are generally made of non-magnetic stainless steel. At the top of the chain of mass comparisons is the world's primary mass standard *defined*[5] as having a mass of 1 kg, which is made of a very dense platinum–iridium alloy and kept (with six copies) at the Bureau International des Poids et Mesures (BIPM) in Paris.[6] Immediately below this level are copies of these standards kept in individual NMIs; the Australian copy ('no. 44') is shown in figure 3.1. Secondary and working mass standards are derived from these and serve the day-to-day needs of scientific research, industry and commerce.

[5] See table 2.1 in section 2.1.2.
[6] See footnote 2 in chapter 3.

The dominant systematic error that arises during weighing is the effect of buoyancy. Since it is not practical to weigh objects in a vacuum, they must be weighed in air. The weight of an object of mass m is then not mg (g being the acceleration due to gravity) but is reduced by the weight of the volume of air that the object displaces.[7] If the object has density ρ, its volume is m/ρ, equal to the volume of displaced air; and if the air has density ρ_a, the weight of this volume is $(m/\rho)\rho_a g = m(\rho_a/\rho)g$. So the weight of the object is not mg but

$$mg - m(\rho_a/\rho)g = mg[1 - (\rho_a/\rho)].$$

If the object is balanced against a mass standard of mass m_s and density ρ_s, we have, on equating weights,

$$m_s g[(1 - (\rho_a/\rho_s)] = mg[1 - (\rho_a/\rho)]. \tag{6.3}$$

In equation (6.3), m_s and m are the 'true' masses of the standard and object, respectively. At the cost of introducing a small systematic error, equation (6.3) may be simplified to an equation directly relating two masses, without any buoyancy terms such as those in square brackets. This simplification makes use of the facts that (as mentioned above) ρ_s is often the density of steel and is therefore near $8\,g \cdot cm^{-3}$; and ρ_a, the density of air, is often near $0.0012\,g \cdot cm^{-3}$ (at $20\,°C$ and near sea-level). These two numerical values are used as standard values in the following definition. Corresponding to m, the 'true mass', we define a 'conventional mass' m_{conv} by the relation

$$m_{conv}[1 - (0.0012/8)] = m[1 - (0.0012/\rho)], \tag{6.4}$$

where the density, ρ, is expressed in $g \cdot cm^{-3}$.

Equation (6.4) states essentially that, for every true mass m of arbitrary density ρ, a conventional mass can be defined as the mass of steel that balances it in air. It is not necessary to specify here whether the mass of steel is the true or conventional mass of steel, because equation (6.4) shows that for a steel object these are equal. Equation (6.4) may be written as

$$m_{conv} = \frac{m[1 - (0.0012/\rho)]}{1 - (0.0012/8)} = \frac{m(\rho - 0.0012)}{0.99985\rho}. \tag{6.5}$$

It is easily checked that, if an object is made of denser material than steel, equation (6.5) implies that its conventional mass will be greater than its true mass. If the object is less dense than steel, its conventional mass will be less than its true mass.

---

[7] This is Archimedes' principle; see Young and Freedman (2003).

There is an old brain teaser: which is heavier, a ton of bricks or a ton of hay? Of course the required answer is that they are equally heavy. On the other hand, the notion of conventional mass accords better with our intuition: the conventional mass of the hay is *very* much less than the conventional mass of the bricks!

With $\rho_a \simeq 0.0012\,\text{g}\cdot\text{cm}^{-3}$ as the density of air and $\rho_s \simeq 8\,\text{g}\cdot\text{cm}^{-3}$ as the density of steel, equations (6.3) and (6.4) together give

$$m_{\text{conv}} = m_s. \tag{6.6}$$

In equation (6.6), m_s is the mass of the steel object of density $\rho_s \simeq 8\,\text{g}\cdot\text{cm}^{-3}$. Equation (6.6) restates equation (6.4): the conventional mass of an object is equal to the mass of the steel standard that balances it in air. The simplicity of this relationship compensates adequately, in most cases, for the generally small systematic error that it introduces.

The proportional systematic error, δ, in the mass of an object which is introduced by the use of the conventional mass is, from equation (6.5),

$$\delta = \frac{(m_{\text{conv}} - m)}{m} = \frac{0.0012(\rho - 8)}{8\rho} = 0.000\,15[1 - (8/\rho)]. \tag{6.7}$$

For an aluminium object, with $\rho \simeq 2.7\,\text{g}\cdot\text{cm}^{-3}$, equation (6.7) implies that there is a systematic error of $\delta = -294$ parts per million in its reported mass.

Exercise D
Find the proportional error for a brass standard of density $8.6\,\text{g}\cdot\text{cm}^{-3}$.

6.1.5 Some sources of systematic errors in temperature measurement

Temperature is an important quantity that is controlled or measured in many experiments. This is due to the fact that many processes and quantities, such as the rates of chemical reactions and the specific heat capacities of solids, are temperature-dependent. Accurate temperature measurement is a challenging pursuit, and many sources of systematic error lie in wait to trap the unwary. If we consider a familiar situation in which the temperature of fluid in a vessel (for example, water) is measured in an open laboratory using a liquid-in-glass thermometer, then the immersed length of the thermometer affects the temperature indicated by the thermometer. The sign of the systematic error depends on whether the fluid is at a higher or lower temperature than the ambient temperature of the laboratory. The magnitude of the error depends on several quantities, including the immersed length and the effective diameter of the thermometer (Nicholas and White 2001). If the temperature of the fluid is changing, the time constant of the thermometer may introduce a significant systematic error. The time constant of a liquid-in-glass thermometer depends on several quantities: the diameter of the thermometer bulb, the heat capacity of the liquid in the thermometer and the heat-transfer coefficient for heat transfer between the thermometer and the fluid in which it is immersed. In situations in which the time constant of a thermometer has been established, it is possible to apply a correction to temperature values to account for the time constant. For example, where a thermometer of time constant τ is used to measure the temperature of a water bath where the rate of temperature rise in the bath is constant at R °C/s, then the lag error L, is given by

$$L = -\tau R. \tag{6.8}$$

In order to account for the finite time constant of the thermometer, the value indicated by the thermometer should be corrected by an amount $+\tau R$.

Exercise E

A liquid-in-glass thermometer with a time constant of 8 s is used to measure the temperature of a water bath. The rate of temperature rise of the water bath is 2.5 °C/minute. At a particular moment, the thermometer indicates that the temperature of the water is 52.5 °C. What is the corrected value for the water temperature which accounts for the time constant of the thermometer?

The extent of the influence of sources of systematic error on measured values of temperature can often be established by changing the conditions of the experiment, and, in general, changing the conditions is an effective means of detecting the existence of a systematic error.

6.2 Systematic error revealed by changed conditions

Accurate measurements are generally made using *formally prescribed* methods.[8] However, there exists a possible risk when prescribed methods are used for prolonged periods without variation: a systematic error may exist or develop, yet remain unsuspected. An effective way of uncovering (and, therefore, correcting for) a systematic error is to *vary* the method by means of an intentional change that does not immediately entail a reduction in accuracy.[9]

---

[8] These are prescribed by a country's *accreditation* bodies, which after inspection of a laboratory and its performance have the power to grant it accreditation for a specified category of measurements. In the USA, the National Institute of Standards and Technology (NIST) operates a Voluntary Laboratory Accreditation Program (VLAP); in the UK accreditation is granted by the UK Accreditation Service (UKAS), while in Australia the accrediting body is the National Association of Testing Authorities (NATA).

[9] For example, replacing an instrument with one of lower accuracy is not the kind of change envisaged here.

Quite small changes in the way a measurement is made can reveal the existence of a large systematic error. For example, varying the immersion depth of a liquid-in-glass thermometer used to measure the temperature of water in a beaker can reveal the extent to which the immersion depth affects the temperature indicated by the thermometer. Similarly, when measuring ambient air temperature with the same type of thermometer, a radiation shield placed around the thermometer can indicate the extent to which a hot or cold heat source in the vicinity of the thermometer is losing heat to, or gaining heat from, the thermometer, thereby affecting the value of temperature that it indicates.

We now examine a more detailed example, taken from electrical metrology, of a change in conditions that reveals a systematic error but that also allows at least partial cancellation of this error.

When a constant ('dc') voltage is intended to be measured with high accuracy using a high-quality DMM, two possible sources of systematic error are the zero-offset of the DMM and thermal voltages caused by the Seebeck effect. The zero-offset is a non-zero voltage displayed by the DMM, due to small imperfections in its electronic circuitry, when the voltage between its input terminals is exactly zero. To achieve a good approximation to this zero voltage, the operator short-circuits the input terminals, using a short length of thick copper wire. The resulting voltage between the input terminals is then likely to be of the order of a few tenths of a microvolt. The display of the DMM can then be 'zeroed' using a pushbutton command. Ideally, this means that the zero offset has been exactly compensated for, so that subsequent readings by the DMM will be free of this offset error. In practice, however, the zero offset will change with time and temperature. The change with temperature is related to the second source of systematic error, due to the Seebeck effect.

The Seebeck effect occurs when dissimilar conductors or semiconductors are joined at their ends to form a loop. A temperature-dependent voltage is generated across each of the two junctions. If the junctions are at exactly the same temperature, the voltages will be equal and opposite and there will be zero net voltage around the loop. If the junctions are at different temperatures, a non-zero net voltage results. To minimise these thermal voltages, copper wiring and terminals are used in high-accuracy electrical metrology, and the copper may be plated with gold or silver to inhibit oxidation (since copper oxides generate relatively large thermal voltages relative to copper). Since small temperature differences will exist between neighbouring terminals even in a temperature-controlled laboratory, particularly immediately after the act of connecting wires to terminals, thermal voltages of the order of tenths of a microvolt or less will still exist. The lead-reversal (or lead-swapping) procedure can eliminate some of these error voltages, as follows.

(a)

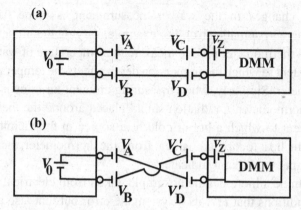

(b)

Figure 6.4. Reversal of connections to eliminate some systematic error voltages.

Figures 6.4(a) and (b) illustrate the voltages just mentioned, in a circuit where a DMM is connected to a source of voltage, V_0. The DMM is represented by a zero-offset voltage, V_Z, at the output of an 'ideal DMM' whose zero-offset voltage is exactly zero. Small circles represent the accessible output terminals of both the DMM and the voltage source. A pair of copper wires (which could in practice be a pair of twin wires within a shielded cable) connects the voltage source to the DMM. Junctions at slightly different temperatures in the internal circuitry of the voltage source produce net thermal voltages V_A and V_B at the external terminals (these voltages may have the opposite sign to that shown). Similarly, at the terminals of the DMM the thermal voltages between the terminals and wires are denoted by V_C and V_D. In this 'forward' measurement (figure 6.3(a)), therefore, the voltage $V_{DMM}^{(f)}$ displayed by the DMM is

$$V_{DMM}^{(f)} = V_0 + V_A - V_B - V_C + V_D - V_Z. \qquad (6.9)$$

There are therefore five unwanted voltages (V_A, V_B, V_C, V_D and V_Z) included in the DMM display. We can eliminate some of them by reversing the leads at the DMM terminals, as shown in figure 6.4(b). Since the connections at the terminals of the voltage source are not touched, the same thermal voltages V_A and V_B are assumed to exist there after the reversal. However, since the connections at the DMM terminals have been changed, and heat has been generated by the act of screwing wires to these terminals, the thermal voltages at the DMM terminals are likely to be different, and are now denoted by V_C' and V_D'. In this 'reverse' measurement, the voltage $V_{DMM}^{(r)}$ displayed by the DMM is

$$V_{DMM}^{(r)} = -V_0 - V_A + V_B - V_C' + V_D' - V_Z. \qquad (6.10)$$

Subtracting equation (6.10) from equation (6.9) gives

$$V_0 = \frac{V_{\mathrm{DMM}}^{(f)} - V_{\mathrm{DMM}}^{(r)}}{2} + (V_B - V_A) + \frac{V_C - V_C'}{2} - \frac{V_D - V_D'}{2}. \tag{6.11}$$

The zero-offset voltage V_Z has therefore been eliminated. If enough time (perhaps a minute or so) has been allowed after reversal of the leads to enable the heat generated by the reversal to dissipate, it is likely that V_C is approximately equal to V_C' and that V_D is approximately equal to V_D'. Then the last two terms in equation (6.11) will be close to zero. The uncancelled thermals are V_A and V_B, but these can be minimised by maintaining the voltage-source terminals as closely as possible at the same temperature. This is sometimes achieved by enclosing the terminals in (but insulating them from) a small metal box.

Lead-swapping as just illustrated is a change in experimental conditions that takes place over seconds or minutes. Other systematic-error-revealing changes in conditions can take place at intervals of months or years. Laboratory conditions are sometimes intentionally changed at intervals for the express purpose of uncovering systematic errors. Regular calibration of all key instruments is an example of an effective change in conditions, and can be conveniently scheduled.

We note also that a systematic error may be uncovered by a change in experimental conditions that occurs 'by accident' or through the passage of time. As noted previously, the intentional exchange of one instrument for another in the same accuracy class is one means of revealing a systematic error. Such an exchange can occur for other reasons. If the readings differ, one or other of the instruments, or both, must be disbelieved. It may then be difficult to trace the history of the systematic error.

The passage of time may create a change in conditions, producing a systematic error that may be significant and yet remain hidden for a prolonged period. For example, the leakage of electric current from a circuit to earth (in practice, the metal enclosure at or close to earth potential in which most electronic instruments are housed) should normally be as low as possible, to correspond to a resistance of about $10^{10}\ \Omega$ or more. This resistance is provided by (among other materials) the plastic insulation around wiring such as polyethylene or polyvinyl chloride. With the passage of time these and other insulators are affected by humidity and absorb various contaminants from the atmosphere. The resulting lower insulation resistance may act like an effective increased loading to a voltage source, creating a significant systematic error.

As an example, in section 6.1.3 a voltage standard of output impedance $1000\ \Omega$ was measured using a DMM with input impedance $10^{10}\ \Omega$. The systematic error was approximately $0.1\ \mu V/V$. Suppose that, over a year, the input impedance gradually decreased to $3 \times 10^9\ \Omega$. The systematic error would increase to about $0.3\ \mu V/V$, but we might incorrectly attribute the reduced reading of the DMM to a real negative drift of the output of the voltage standard.

6.3 Review

In this chapter we have considered several sources of systematic error and how the effect of those errors can be minimised, or accounted for. We have shown that systematic errors can be quantified through Type A or Type B evaluations of uncertainty, or sometimes using a combination of both types of evaluation. Next we consider in more detail how uncertainties are calculated, and how they may be combined.

7

Calculation of uncertainties

Random errors, evaluated using statistical methods, create a Type A uncertainty. A known systematic error in a measured value should be corrected for, and after the correction has been made, the uncertainty in the correction contributes to the uncertainty in that value. The uncertainty in the correction, and hence in the value, may be Type A or Type B, depending on how the uncertainty is evaluated. The finally reported uncertainty of a measurand, called the *combined* uncertainty, is likely to have both Type A and Type B components, but becomes wholly Type B when subsequent use is made of it.

In this chapter we consider how to evaluate the combined uncertainty of a measurand. The procedure to be described makes no distinction between Type A and Type B uncertainties. It may appear then as if we have gone to unnecessary trouble in assigning types to uncertainties, but this classification is desirable since it emphasises the different methods by which they are evaluated. It is also useful as a reminder that, whereas an 'error' can be random or systematic, 'uncertainty' is a separate concept whose two types are distinguished from each other by different names, 'Type A' and 'Type B'. However, once uncertainties have been classified, Type A and Type B uncertainties are treated identically thereafter.

7.1 The measurand model and propagation of uncertainties from inputs to measurand

A measurand, which by definition is the particular quantity to be determined, often cannot be measured directly. Instead, we measure the *input* quantities that determine the value of the measurand.[1]

---

[1] Input quantities are sometimes referred to as *influence* quantities, since they 'influence' the measurand.

If there are n input quantities, $x_1, x_2, \ldots, x_n$, we describe their relationship to the measurand, y, by the functional relationship

$$y = f(x_1, x_2, \ldots, x_n). \tag{7.1}$$

Equation (7.1) is our measurand model. In some situations, $x_1, x_2, \ldots, x_n$ represent values of the same quantity obtained through repeated measurements. In other cases, $x_1, x_2, \ldots, x_n$ represent different types of quantities. For example, in a situation in which y depends on three input quantities, x_1 might represent a length, x_2 a temperature and x_3 a thermal conductivity.

Equation (7.1) is the relationship between the *estimates*, $x_1, x_2, \ldots, x_n$ and the resulting *estimate*, y, of the measurand. This relationship between estimates is the experimentally feasible counterpart to the corresponding relationship usually expressed in upper-case symbols as $Y = f(X_1, X_2, \ldots, X_n)$. Here $X_1, X_2, \ldots, X_n$ are the values ('actual' or 'true' values) of the inputs, and Y is the value ('actual' or 'true' value) of the measurand. There is, therefore, a useful conceptual distinction between 'estimate' (short for 'estimate of value') and 'true value'. However, in practical applications of the propagation formula to be derived below (equation (7.14)), it is convenient to use upper-case or lower-case symbols to represent estimates in accordance with existing notational convention for physical quantities; for example, estimates of volume or voltage will be denoted by upper-case V.

A small change, δy, in y, is related to small changes, $\delta x_1, \delta x_2, \ldots, \delta x_n$, in $x_1, x_2, \ldots, x_n$ respectively, by

$$\delta y = \frac{\partial y}{\partial x_1} \delta x_1 + \frac{\partial y}{\partial x_2} \delta x_2 + \cdots + \frac{\partial y}{\partial x_n} \delta x_n, \tag{7.2}$$

where $\partial y/\partial x_1, \partial y/\partial x_2, \ldots, \partial y/\partial x_n$ are the first-order partial derivatives of y with respect to $x_1, x_2, \ldots, x_n$ respectively.

Equation (7.2) can be seen to be plausible by considering the case of a single input, x, and its effect on y. Figure 7.1 shows the response of y to x, and we now examine the effect of a small change, δx, in x from its initial value, x_0. The point (x_0, y_0) is labelled P in figure 7.1. If δx is small, the response of y is linear. This straight-line portion of the curve near x_0, namely the arc PQ, may be approximated by the equation, $y = A + Bx$, where A and B are constants. The derivative or gradient, dy/dx, at $x = x_0$ is therefore $dy/dx = B$. At $x = x_0$, we have $y = y_0 = A + Bx_0$. When the input, x, changes to $x_0 + \delta x$, y changes to $y_0 + \delta y = A + B(x_0 + \delta x)$. This point, $(x_0 + \delta x, y_0 + \delta y)$, is labelled Q in figure 7.1. Therefore $\delta y = A + B(x_0 + \delta x) - A - Bx_0 = B\delta x = (dy/dx)\delta x$. Equation (7.2) is a generalisation for several inputs, x_i, of this linear approximation of the response of the measurand to its inputs.

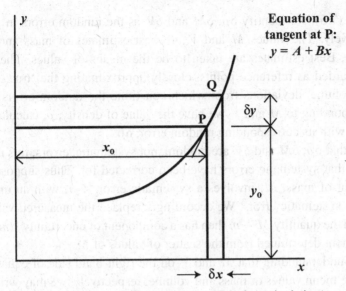

Figure 7.1. Demonstration of equation (7.2) for a single input x.

As an example of the application of equation (7.2), we may wish to determine the density, ρ, of an object of mass M and volume V. Here the measurand is ρ, and the two input quantities are M and V. The relationship between the measurand and the input quantities (sometimes called the measurand model) is

$$\rho = \frac{M}{V}. \tag{7.3}$$

Since neither M nor V can be known exactly, each must have an associated uncertainty. It follows that ρ will also have an uncertainty. We speak of the uncertainties in the inputs M and V as 'propagating' into ρ and causing a corresponding uncertainty in ρ. To see in detail how uncertainties propagate, we consider the following argument and keep in mind that, while error may be positive or negative, uncertainty is a positive quantity.

If $\rho = M/V$, then the differential, $\delta\rho$, represents a small increase or decrease in ρ. Using equation (7.2), $\delta\rho$ is given by

$$\delta\rho = \frac{\partial\rho}{\partial M}\,\delta M + \frac{\partial\rho}{\partial V}\,\delta V. \tag{7.4}$$

Since (using equation (7.3)) $\partial\rho/\partial M = 1/V$ and $\partial\rho/\partial V = -M/V^2$, equation (7.4) becomes[2]

$$\delta\rho = \frac{1}{V}\,\delta M - \frac{M}{V^2}\,\delta V. \tag{7.5}$$

---

[2] It is helpful to check the dimensional consistency of all the terms in the expression for the differential, so that any mistake in the expression can be identified and corrected.

In equation (7.5), we identify $\delta\rho$, δM and δV as the random errors in ρ, M and V, respectively. The values M and V are best estimates of mass and volume, respectively. Best estimates are taken to be the means of values. These means may be regarded as reference points, closely approximating the 'true' values of mass and volume, deviations from which constitute the random errors, δM and δV. Corresponding to M and V we have the value of density, ρ, calculated using $\rho = M/V$, with its corresponding random error, $\delta\rho$.

We note that $\delta\rho$, δM and δV are random, not systematic, errors; this is because we assume that systematic errors have been corrected for. Thus suppose that the measurement of mass, M, involves a systematic error, $+m$, with an uncertainty $s(m)$ in this systematic error.[3] We accordingly replace the measured value M by $M - m$, and the quantity $M - m$ then has a component of uncertainty $s(m)$, as well as a component determined from the scatter of values of M.

With the understanding that M and V on the right-hand side of equation (7.5) represent the mean values of mass and volume, respectively, we may write

$$\delta M = M_k - M. \tag{7.6}$$

The index, k, on the right-hand side of equation (7.6) expresses explicitly the fact that δM is not a single random error but represents a set of N random errors, where N is the very large or infinite population of random errors.[4] Thus $k = 1, 2, \ldots, N$. Similarly, $\delta V = V_k - V$ and $\delta\rho = \rho_k - \rho$. In all three cases we may consider k as running from 1 to the same very large or infinite number, N. The quantity ρ represents the mean value of density.[5]

Equation (7.5) may now be written, for *each* random error $\rho_k - \rho$, $M_k - M$ and $V_k - V$, as

$$\rho_k - \rho = \frac{1}{V}(M_k - M) - \frac{M}{V^2}(V_k - V). \tag{7.7}$$

If we sum equation (7.7) over all k (from $k = 1$ to $k = N$),

$$\sum_{k=1}^{N}(\rho_k - \rho) = \frac{1}{V}\sum_{k=1}^{N}(M_k - M) - \frac{M}{V^2}\sum_{k=1}^{N}(V_k - V), \tag{7.8}$$

we obtain zero for each of the three terms, since $\sum_{k=1}^{N}\rho_k = N\rho$, $\sum_{k=1}^{N}M_k = NM$ and $\sum_{k=1}^{N}V_k = NV$ and ρ, M and V are the mean values of density, mass and volume.

---

[3] As in chapter 6, we take the correction for a systematic error to have the opposite sign to the systematic error.
[4] In a real experiment, we sample the population by making n measurements, where $n \ll N$.
[5] We note that, if the errors δM occur independently of the errors δV, then (following rule (d) in section 5.1.1), we have $\rho = M/V$ as the relation obeyed by the mean values of ρ, M and V.

If we square each side of equation (7.7), we have

$$(\rho_k - \rho)^2 = \frac{1}{V^2}(M_k - M)^2 + \frac{M^2}{V^4}(V_k - V)^2 - \frac{2M}{V^3}(M_k - M)(V_k - V), \quad (7.9)$$

and summing over all k from 1 to N and dividing by N gives

$$\frac{\sum_{k=1}^{N}(\rho_k - \rho)^2}{N} = \frac{1}{V^2}\frac{\sum_{k=1}^{N}(M_k - M)^2}{N} + \frac{M^2}{V^4}\frac{\sum_{k=1}^{N}(V_k - V)^2}{N}$$
$$- \frac{2M}{V^3}\frac{\sum_{k=1}^{N}[(M_k - M)(V_k - V)]}{N}. \quad (7.10)$$

The term on the left-hand side of equation (7.10) is the variance of the density in its population or, equivalently, the squared standard uncertainty, $u^2(\rho)$, of the density. Similarly we have

$$u^2(M) = \frac{\sum_{k=1}^{N}(M_k - M)^2}{N},$$
$$u^2(V) = \frac{\sum_{k=1}^{N}(V_k - V)^2}{N}. \quad (7.11)$$

We now examine the third term on the right-hand side of equation (7.10). The errors, $\delta M = M_k - M$ and $\delta V = V_k - V$, are unlikely to exhibit any degree of mutual dependence, since M and V are measured using completely different instruments (for example, scales for the mass, M, and vernier calipers for the length measurements used to determine the volume, V). We say that the measurements of M and of V are likely to be *uncorrelated*; a positive error δM is likely to be associated just as often with a negative error δV as with a positive error δV. Similarly, a negative error δM is likely to be associated just as often with a positive error δV as with a negative error δV. This implies that the product of each instance of error $\delta M = M_k - M$ with a corresponding instance of error $\delta V = V_k - V$ will be zero on average. The quantity $\sum_{k=1}^{N}[(M_k - M)(V_k - V)]$ is therefore zero. Equation (7.10) simplifies to

$$u^2(\rho) = \frac{1}{V^2}u^2(M) + \frac{M^2}{V^4}u^2(V). \quad (7.12)$$

In metrology, the notation $u^2(x)$ is most commonly used for the variance of a quantity, x, in its population, and $u(x)$ denotes the square root of the variance, namely the standard deviation or standard uncertainty of x.

The law of propagation of uncertainties as expressed by equation (7.12) is written in terms of variances, namely the squared standard uncertainties. The variance of a quantity has the dimensions of that quantity squared; thus $u^2(\rho)$ has the dimensions of density squared, and it may be verified that the three terms in equation (7.12) have the same dimensions. From equation (7.12) it follows that the combined standard uncertainty, $u(\rho)$, in ρ is given by

$$u(\rho) = \sqrt{\frac{1}{V^2}u^2(M) + \frac{M^2}{V^4}u^2(V)}.$$

(7.13)

Suppose that $V = 3.930\,\text{cm}^3$ and $M = 10.601\,\text{g}$, so that $\rho = 2.697\,\text{g}\cdot\text{cm}^{-3}$. If the standard uncertainty, $u(V)$, in V is $0.002\,\text{cm}^3$ and the standard uncertainty, $u(M)$, in M is 5 mg, it may be confirmed that $u(\rho) \simeq 1.9\,\text{mg}\cdot\text{cm}^{-3}$.

We should note that inputs may exhibit a mutual non-zero correlation. In the particular case above, suppose that δM and δV did have some degree of mutual dependence. For example, if, whenever δM was positive, so was δV, and whenever δM was negative, so was δV, then M and V would exhibit mutual positive correlation. The summation in the third term on the right-hand side of equation (7.10) would then give a positive result, and (because of the minus sign in front) that third term would be negative. Similarly, δM and δV would exhibit mutual negative correlation if, whenever δM was positive, δV was negative, and whenever δM was negative, δV was positive. The third term on the right-hand side in equation (7.10) would then be positive. Correlations between inputs will be considered further in section 7.2.

When y depends on an arbitrary number of input quantities, as expressed by equation (7.1), the uncertainties $u(x_i)$ ($i = 1, 2, \ldots, n$) propagate into y according to

$$u^2(y) = \left(\frac{\partial y}{\partial x_1}\right)^2 u^2(x_1) + \left(\frac{\partial y}{\partial x_2}\right)^2 u^2(x_2) + \cdots + \left(\frac{\partial y}{\partial x_n}\right)^2 u^2(x_n)$$

(7.14)

provided that the x_i ($i = 1, 2, \ldots, n$) are mutually uncorrelated.

Exercise A
(1) The frequency, f, of a waveform is related to the period, T, of the waveform by the relationship

$$f = \frac{1}{T}.$$

Given that $T = 21.5\,\text{ms}$ and $u(T) = 2.4\,\text{ms}$, calculate f and $u(f)$.
(2) The gain, G, of a non-inverting amplifier is expressed as a ratio of two resistances, R_1 and R_2, given by

$$G = 1 + \frac{R_2}{R_1}.$$

If $R_1 = 1053\,\Omega$, $u(R_1) = 12\,\Omega$, $R_2 = 12\,350\,\Omega$ and $u(R_2) = 95\,\Omega$, calculate the gain of the amplifier and the standard uncertainty in the gain.

(3) The velocity, v, of a wave on a stretched string may be written as

$$v = \sqrt{\frac{T}{\mu}},$$

where T is the tension in the string and μ is the mass per unit length of the string. Assume that $T = 2.51\,\text{N}$, $u(T) = 0.05\,\text{N}$, $\mu = 1.032\,\text{g/m}$ and $u(\mu) = 0.012\,\text{g/m}$. Determine

(a) expressions for $\partial v/\partial T$ and $\partial v/\partial \mu$, and

(b) the velocity of the wave and the standard uncertainty in the velocity.

(4) The focal length, f, of a thin lens is related to the distance, p, from an object to the lens, and the distance, q, from the image to the lens, by the relationship

$$\frac{1}{f} = \frac{1}{p} + \frac{1}{q}.$$

Assume that $p = 12.5\,\text{cm}$, $u(p) = 0.5\,\text{cm}$, $q = 42.5\,\text{cm}$ and $u(q) = 1.5\,\text{cm}$. Determine

(a) expressions for $\partial f/\partial p$ and $\partial f/\partial q$, and

(b) f and $u(f)$.

7.1.1 Sensitivity coefficients

The partial derivatives in equation (7.14) are sometimes called *sensitivity coefficients*, and are represented by the symbol, c. Thus the degree of sensitivity, $\partial y/\partial x_1$, of y to x_1, in equation (7.14), may be called c_1 (so that the coefficient of $u^2(x_1)$ in equation (7.14) is c_1^2). The c notation is useful as a shorthand when a table of uncertainty contributions from various inputs is being drawn up.

If the measurand is the sum of the inputs,

$$y = x_1 + x_2 + \cdots + x_n, \tag{7.15}$$

then $c_i = \partial y/\partial x_i = 1$ for all i ($i = 1, 2, \ldots, n$), and equation (7.14) gives

$$u^2(y) = u^2(x_1) + u^2(x_2) + \cdots + u^2(x_n)$$

or

$$u(y) = \sqrt{u^2(x_1) + u^2(x_2) + \cdots + u^2(x_n)}. \tag{7.16}$$

Equation (7.16) shows that $u(y)$ is the 'root-sum-square' of the $u(x)$'s. Combining standard uncertainties, whether Type A or Type B, by root-sum-squares is the correct procedure when the x's (or, more precisely, their errors) are uncorrelated. This contrasts with the past (and no longer recommended) practice of simply adding the uncertainties, which pessimistically gives a larger $u(y)$ and neglects the fact that uncorrelated errors are likely to exert some degree of mutual cancellation.

If the measurand, y, is proportional to a single input, x, so that

$$y = Kx, \tag{7.17}$$

where K is a constant, we have $c = \partial y / \partial x = K$, and equation (7.14) gives[6]

$$u^2(y) = K^2 u^2(x). \tag{7.18}$$

Bearing in mind the above definition of the sensitivity coefficients, c_i, as $c_i = \partial y / \partial x_i$, we see that equation (7.14) may be written in a form that is a generalisation of equation (7.16):

$$u(y) = \sqrt{c_1^2 u^2(x_1) + c_2^2 u^2(x_2) + \cdots + c_n^2 u^2(x_n)}. \tag{7.19}$$

Equation (7.19) shows, essentially, that all the Type A standard uncertainties can be combined by root-sum-squares to give a 'net' Type A component, and similarly for all the Type B components. We now have the prescription for obtaining the combined standard uncertainty due to several inputs: it is the root-sum-square of the Type A and Type B components.

Exercise B
For the following equations, determine the sensitivity coefficients, $c_1 = \partial y / \partial x_1$, $c_2 = \partial y / \partial x_2$, etc.

(a)

$$y = \frac{x_1^2 x_2}{x_3}.$$

(b)

$$y = \sqrt{\frac{x_1}{2x_2}}.$$

(c)

$$y = x_1 \exp x_2.$$

(d)

$$y = \frac{\sin x_1}{\sin x_2}.$$

---

[6] Equations (7.17) and (7.18) are obtained here in the context of uncertainty in measurement, but they also constitute an elementary but fundamental result in statistics. If two variables x and y are related by $y = Kx$, with K constant, then the variance of y is $K^2 \times$ variance of x.

7.1.2 Use of least-squares with the measurand model

To be able to apply the measurand model given by equation (7.1) and leading to the propagation equation (7.14), we need the best estimate of each input quantity and the standard uncertainty in that estimate. Very often, the technique of least-squares is used to establish best estimates of input quantities and of their associated standard uncertainties. Owing to the similarity in the nomenclature used, it is quite easy to confuse the x's used in the measurand model and those used when applying least-squares.

In the measurand model (equation (7.1)), x_i represents the best estimate of the ith input quantity and $u(x_i)$ is its standard uncertainty to be inserted into equation (7.14). Each x_i may represent a different physical quantity with different dimensions. By contrast, x_i in ordinary least-squares normally denotes the ith value of the predictor (or 'explanatory') variable. All the x_i in the ordinary least-squares model are values of the same physical quantity with the same dimensions, and they are all assumed to be error-free and therefore to have no uncertainty. It is the *parameters* within the least-squares model (such as the mean, or slope and intercept) that are estimated and these, together with their associated standard uncertainties, become inputs to the measurand model.

As the simplest and very common example, one (or more) of the x_i in the measurand model might be the *mean* of several values obtained through repeated readings. As discussed in section 5.2.1, the calculation of the mean is the simplest case of a least-squares fit. Thus an input, x_1 (for example), in the measurand model would then be calculated as

$$x_1 = (x_{11} + x_{12} + \cdots + x_{1q})/q,$$

where $x_{11}, x_{12}, \ldots, x_{1q}$ are the q values for the first input, x_1. If these values have an unbiased variance, s^2, calculated in the usual way as

$$s^2 = \frac{\sum_{i=1}^{q}(x_{1i} - x_1)^2}{q - 1},$$

and if the readings are uncorrelated, we have

$$u(x_1) = \frac{s}{\sqrt{q}}, \tag{7.20}$$

which restates equation (5.56). The squared value, $u^2(x_1) = s^2/q$, is then the correct entry on the right-hand side of equation (7.14).

Similarly, one of the inputs may be the estimated value, b, of a drift in time, determined by a least-squares fit of a response variable to q points in time t_1, $t_2, \ldots, t_q$ as in section 5.2.3. Then s_b given by equation (5.58) is expressed in the new notation as $u(b)$, given by

$$u(b) = s\sqrt{\frac{q}{D}}, \tag{7.21}$$

where s is the rms residual about this line and D is given by $D = q\sum_{i=1}^{q} t_i^2 - \left(\sum_{i=1}^{q} t_i\right)^2$ (see equation (5.51)). The squared value, $u^2(b)$, is then the correct entry for the squared standard uncertainty of the drift input on the right-hand side of equation (7.14).

In general, least-squares (including the simple case of calculating a mean) is the technique by which we estimate our Type A uncertainties of the inputs on the right-hand side of equation (7.14). Type B uncertainties, which are not evaluated using statistical analysis, refer to single values of the inputs, since repeated readings are usually not available. However, the single value is nevertheless the mean in the sense of a best estimate. This is why, in the example leading to equation (7.12), M and V were called the mean mass and volume, respectively, with $u(M)$ and $u(V)$ as the measures of the uncertainties in these means created by the population of errors, $M_k - M$ and $V_k - V$.

Example 1
A current, I, is calculated using Ohm's Law: $I = V/R$. V is the measured value of voltage. The resistance, R, is not measured directly, but is found with the assistance of the temperature coefficient of resistance of R, as obtained from a calibration report. Specifically, R is found using the relationship

$$R = R_0 + \alpha(t - t_0), \tag{7.22}$$

where R_0 is the resistance at a fixed reference temperature, t_0, and α is the usual symbol for the temperature coefficient[7] at t_0. We can measure the temperature, t, with standard uncertainty, $u(t)$. The calibration report states R_0, $u(R_0)$, α, $u(\alpha)$ and t_0. From equation (7.22) we have

$$\frac{\partial R}{\partial R_0} = 1, \tag{7.23}$$

$$\frac{\partial R}{\partial \alpha} = (t - t_0), \tag{7.24}$$

$$\frac{\partial R}{\partial t} = \alpha, \tag{7.25}$$

so that

$$u^2(R) = u^2(R_0) + (t - t_0)^2 u^2(\alpha) + \alpha^2 u^2(t). \tag{7.26}$$

---

[7] The form of equation (7.22) implies that α has units of ohms per degree. This simplifies the following analysis, but in practice α is more likely to be given in proportional parts of resistance per degree – for example, $(\mu\Omega/\Omega)\cdot{}^{\circ}\mathrm{C}^{-1}$.

Equation (7.14) becomes

$$u^2(I) = \frac{1}{R^2}u^2(V) + \frac{V^2}{R^4}[u^2(R_0) + (t - t_0)^2 u^2(\alpha) + \alpha^2 u^2(t)], \qquad (7.27)$$

where $R = R_0 + \alpha(t - t_0)$

The standard uncertainties, $u(R_0)$ and $u(\alpha)$, in the calibration report are likely to have been determined from a linear least-squares fit similar to that described in section 5.2.3.[8]

In this example, V is measured using a DMM. If several repeated measurements are made of V, the standard deviation of the mean voltage is the Type A component in $u(V)$. The Type B component is, for example, the standard uncertainty of the correction to be applied to the readings of the DMM.[9] The standard uncertainty, $u(V)$, is the root-sum-square of the Type A and Type B components.

The other terms in equation (7.27) are also known: the values of $R_0, \alpha, t_0, u(R_0)$ and $u(\alpha)$ are stated in the calibration report on the resistor, and we can measure the temperature, t, of the resistor at the time of the experiment.

Exercise C
Assume that $V = 1.32$ V, $u(V) = 0.02$ V, $R_0 = 1032\,\Omega$, $u(R_0) = 23\,\Omega$, $t = 32.5\,°C$, $u(t) = 0.5\,°C$, $t_0 = 25\,°C$, $\alpha = 4.35\,\Omega/°C$ and $u(\alpha) = 0.03\,\Omega/°C$.

Use $I = V/R$ together with equations (7.22) and (7.27) to calculate R, I and $u(I)$.

Example 2
Equation (7.14) can be applied to the common case where the inputs, x_i ($i = 1, 2, \ldots, n$), are n repeated and mutually uncorrelated values of the same quantity, and the measurand, y, is the mean of the inputs:

$$y = \frac{x_1 + x_2 + \cdots + x_n}{n}. \qquad (7.28)$$

Since $\partial y/\partial x_i = 1/n$ for all i, equation (7.14) gives

$$u^2(y) = \frac{1}{n^2}[(u^2(x_1) + u^2(x_2) + \cdots + u^2(x_n)]. \qquad (7.29)$$

---

[8] By rewriting the relationship $R = R_0 + \alpha(t - t_0)$ in the form $R = R_0 + \alpha t'$, where t' is defined as the deviation from the fixed and known temperature t_0, we see that the relationship between R and t' is the same as that between voltage, V, and time, t, in equation (5.35). Equations (5.57) and (5.58) give the standard uncertainties, s_a and s_b, of intercept and slope, respectively, and these are equivalent to $u(R_0)$ and $u(\alpha)$, respectively, in this example.

[9] This correction, which itself depends on the voltage, and its standard uncertainty are usually available from the calibration report on that DMM.

Table 7.1. *Thickness of*
aluminium film

Thickness (nm)
320
330
315
330
325
315

We need an estimate of each $u^2(x_i)$. An estimate of *each* $u^2(x_i)$ is the variance, s^2, as calculated using equation (5.23). It follows that all the $u^2(x_i)$ on the right-hand side of equation (7.29) are equal and we write $u^2(x_i) = s^2 = u^2(x)$. So equation (7.29) gives

$$u^2(y) = \frac{1}{n^2}[nu^2(x)] = \frac{u^2(x)}{n} \tag{7.30}$$

or

$$u(y) = \frac{u(x)}{\sqrt{n}}. \tag{7.31}$$

This result in a different notation was stated in equations (4.3) and (5.56).

Exercise D
The thickness of a thin film of aluminium deposited onto a glass slide is measured at several points using a profilometer. The values obtained are shown in table 7.1.

Calculate the mean of the values in table 7.1 and the standard uncertainty in the mean, assuming that measurement errors are uncorrelated.

Example 3
Suppose that each input, x_i, to a measurand is the mean of n_i values obtained by repeat measurement. We write the n_i values as $x_{i1}, x_{i2}, \ldots, x_{in_i}$. The mean, x_i, of the ith input is given by $x_i = (x_{i1} + x_{i2} + \cdots + x_{in_i})/n_i$, and the standard deviation, s_i, of these n_i values is

$$s_i = \sqrt{\frac{(x_{i1} - x_i)^2 + (x_{i2} - x_i)^2 + \cdots + (x_{in_i} - x_i)^2}{n_i - 1}}. \tag{7.32}$$

Table 7.2. *Focal lengths of objective and eyepiece lenses*

f_o (cm)	f_e (cm)
30.3	5.6
30.7	5.5
30.5	5.2
30.6	5.5
31.1	5.4
30.2	
30.4	
30.4	

If the n_i values are uncorrelated, the standard deviation of the mean, x_i, is given by $s_i / \sqrt{n_i}$. In the notation of the measurand model, we have $u(x_i) = s_i / \sqrt{n_i}$.

Exercise E

The magnification, m, of a refracting telescope is equal to the ratio of the focal length of the lenses in the telescope:

$$m = \frac{f_o}{f_e},$$

where f_o is the focal length of the objective lens and f_e is the focal length of the eyepiece lens. Repeat measurements of the focal length of each lens are made. These are shown in table 7.2.

Use the data in table 7.2 to find

(a) the mean focal length of each lens;
(b) the standard uncertainty in the mean focal length of each lens;
(c) the best estimate of the magnification of the telescope; and
(d) the combined standard uncertainty in the magnification, assuming that errors in f_o and f_e are uncorrelated.

7.2 Correlated inputs

The expression $\sum_{k=1}^{N} [(M_k - M)(V_k - V)]/N$, in the third term on the right-hand side of equation (7.10), was assumed to be zero, expressing the lack of mutual correlation between the errors, $M_k - M$ and $V_k - V$. This expression may be recognised as the covariance of M and V. Just as $u^2(x)$ denotes the variance of x, a convenient symbol for the covariance of two quantities, x and y, is $u(x, y)$. We may write that covariance as $u(M, V)$, in which case equation (7.10) becomes[10]

$$u^2(\rho) = \frac{1}{V^2}u^2(M) + \frac{M^2}{V^4}u^2(V) - \frac{2M}{V^3}u(M, V). \tag{7.33}$$

The standard uncertainty, $u(x)$, of x, is the square root of the variance. Expressed in our present notation, the correlation coefficient, r, between variables x and y, is[11]

$$r(x, y) = \frac{u(x, y)}{u(x)u(y)}. \tag{7.34}$$

Equation (7.33) can now be written

$$u^2(\rho) = \frac{1}{V^2}u^2(M) + \frac{M^2}{V^4}u^2(V) - \frac{2M}{V^3}r(M, V)u(M)u(V), \tag{7.35}$$

where $r(M, V)$ denotes the correlation coefficient between M and V. We note that, if small changes in variables (like M and V) are being considered and these variables are said to be correlated, this is equivalent to saying that the *errors* (like $\delta M = M_k - M$ and $\delta V = V_k - V$) in those variables are correlated.

For correlated inputs, equation (7.35) suggests that the general form of equation (7.14) should be

$$u^2(y) = \left(\frac{\partial y}{\partial x_1}\right)^2 u^2(x_1) + \left(\frac{\partial y}{\partial x_2}\right)^2 u^2(x_2) + \cdots + \left(\frac{\partial y}{\partial x_n}\right)^2 u^2(x_n)$$

$$+ r(x_1, x_2)\frac{\partial y}{\partial x_1}\frac{\partial y}{\partial x_2}u(x_1)u(x_2) + r(x_1, x_3)\frac{\partial y}{\partial x_1}\frac{\partial y}{\partial x_3}u(x_1)u(x_3) + \cdots$$

$$+ r(x_i, x_j)\frac{\partial y}{\partial x_i}\frac{\partial y}{\partial x_j}u(x_i)u(x_j) + \cdots, \tag{7.36}$$

where $r(x_i, x_j)$ is the correlation coefficient between inputs x_i and x_j.

There are $n(n-1)$ 'product' terms in equation (7.36). This can be seen by noting that

x_1 is associated with the $n-1$ other terms $x_2, x_3, \ldots, x_n$;
x_2 is associated with the $n-1$ other terms $x_1, x_3, x_4, \ldots, x_n$;

and so on for all n terms, each being associated with the product of the $n-1$ *other* terms. Since, for example, $r(x_1, x_2)(\partial y/\partial x_1)(\partial y/\partial x_2)u(x_1)u(x_2) =$

---

[10] The dimensions of a covariance such as $u(M, V)$ should be noted: they are, in this case, mass × volume, so in general the dimensions of the covariance, $u(x, y)$, are the product of the dimensions of x and of y.
[11] See equation (5.60) for a definition of r.

$r(x_2, x_1)(\partial y/\partial x_2)(\partial y/\partial x_1)u(x_2)u(x_1)$, the $n(n-1)$ product terms come in $\frac{1}{2}n(n-1)$ pairs, in each of which the two terms are identical. In equation (7.10), for example, the coefficient $-2M/V^3$ of the third term on the right-hand side is the sum

$$\frac{\partial \rho}{\partial M}\frac{\partial \rho}{\partial V} + \frac{\partial \rho}{\partial V}\frac{\partial \rho}{\partial M}.$$

It follows that equation (7.36) may be written

$$u^2(y) = \left(\frac{\partial y}{\partial x_1}\right)^2 u^2(x_1) + \left(\frac{\partial y}{\partial x_2}\right)^2 u^2(x_2) + \cdots + \left(\frac{\partial y}{\partial x_n}\right)^2 u^2(x_n)$$

$$+ 2r(x_1, x_2)\frac{\partial y}{\partial x_1}\frac{\partial y}{\partial x_2}u(x_1)u(x_2) + 2r(x_1, x_3)\frac{\partial y}{\partial x_1}\frac{\partial y}{\partial x_3}u(x_1)u(x_3) + \cdots$$

$$+ 2r(x_i, x_j)\frac{\partial y}{\partial x_i}\frac{\partial y}{\partial x_j}u(x_i)u(x_j) + \cdots, \tag{7.37}$$

where now the second suffix, j, is always greater than the first suffix, i.

When the measurand, y, is the mean of uncorrelated inputs (such that $r(x_i, x_j) = 0$ for $i \neq j$) obtained as a time-sequence of repeated readings $x_1, x_2, \ldots, x_n$, we have the result $u(y) = u(x)/\sqrt{n}$, as in equation (7.31). We now consider how this result is modified when (for example) all the x_i's are perfectly mutually correlated, with a correlation coefficient of $+1$.

7.2.1 Increase in uncertainty in the measurand due to correlated inputs

A correlation coefficient between two populations is defined through a one-to-one correspondence between their respective elements. A high positive correlation between them exists when high values in one population are associated with high values in the other or when low values in one population are associated with low values in the other.

For the particular case of repeated readings of the same quantity, it is not immediately obvious how two such populations can arise when we have only a single sequence of values obtained by repeat measurements. Unless mentioned otherwise, we shall assume that the sequence is a time-sequence, its terms having been obtained at successive instants of time separated by equal intervals. The two populations are generated conceptually by regarding the single actual sequence as representative of many possible sequences. The two populations are, then, the populations formed by (for example) the first and second terms (or any pair of terms) in each of the possible sequences. It is essential to regard a sequence (whether actual or possible) as *ordered*; its terms cannot be shuffled.

An example of high positive correlation between any two of the n inputs x_1, $x_2, \ldots, x_n$ occurs if they constitute a set of values, obtained at equal time intervals, of the same quantity that exhibits a steady drift in time. Thus suppose that, because of a steady drift in time, our n inputs have the sequentially obtained values (for simplicity) $x_1 = 1, x_2 = 2, \ldots, x_n = n$. We now imagine that we immediately take another sample of n values – in other words, a second actual sequence – and (because we assume that the same drift still exists) we now obtain $n + 1, n + 2, \ldots, 2n$. A third sample gives $2n + 1, 2n + 2, \ldots, 3n$. So, if we draw up columns of (say) first and second inputs in each round, the entries will look like this:

$$
\begin{array}{cc}
1 & 2 \\
n + 1 & n + 2 \\
2n + 1 & 2n + 2 \\
3n + 1 & 3n + 2 \\
\cdots & \cdots
\end{array}
$$

exhibiting perfect positive correlation between the two inputs. The same perfect positive correlation exists between the first and third inputs, between the second and third inputs, and indeed between any pair of inputs. This imaginary exercise shows that, in our *single* actual sequence with a steady drift, the values have perfect mutual positive correlation ($r = +1$).

We now calculate the mean, y, of the inputs obtained as our single set of n repeated readings:

$$
y = \frac{x_1 + x_2 + \cdots + x_n}{n}, \tag{7.38}
$$

so that $\partial y/\partial x_i = 1/n$ for all i. With $r(x_1, x_2) = +1$ and all the $u(x_1) = u(x_2) = \cdots = u(x_n) = u(x)$, equation (7.36) gives

$$
u^2(y) = \frac{1}{n^2}[nu^2(x)] + 1\frac{1}{n}\frac{1}{n}n(n - 1)u^2(x). \tag{7.39}
$$

In the second term in equation (7.39), $+1$ is the correlation coefficient, the next two factors, each $1/n$, are (as shown immediately after equation (7.38)) the two partial derivatives for the product terms in equation (7.36), and the factor $n(n - 1)$ is present because there are $n(n - 1)$ such identical product terms in equation (7.36). Equation (7.39) therefore gives

$$
u^2(y) = u^2(x)\left(\frac{1}{n} + \frac{n - 1}{n}\right) = u^2(x). \tag{7.40}
$$

We conclude that the standard deviation of the mean remains the same as the standard deviation of the distribution of the x values.

7.2.2 *The experimental standard deviation of the mean (ESDM) and the divisor* $\sqrt{n}$

Equation (7.40), which can be written $u(y) = u(x)$ and applies to the case of perfectly correlated readings, contrasts with equation (7.31), $u(y) = u(x)/\sqrt{n}$, for uncorrelated readings. The standard uncertainty, $u(y)$, of the mean of repeated values is often called the experimental standard deviation of the mean (ESDM).[12] In this section we discuss the validity of the formula $u(y) = u(x)/\sqrt{n}$ for the ESDM. We describe, in general terms, some of the tools available for treating those cases where, because of correlations, the ESDM is not derived from the standard deviation simply by dividing by $\sqrt{n}$.

Although perfect correlation is rarely seen, nevertheless, if repeated readings exhibit a significant drift in time, we should be cautious about claiming that the uncertainty of the mean is reduced by a factor of $\sqrt{n}$ compared with the uncertainty of the individual values. Ideally we should take the drift into account, by fitting a straight line to data using least-squares. If this is not practicable, we should state $u(y) = u(x)$ as implied by equation (7.40), so that the standard uncertainty in the mean is simply the standard deviation of the values. This non-reduction in uncertainty is intuitively acceptable for this case of drift, if we remember that the purpose of taking repeated readings is to cancel out random errors.[13] However, a drift that gives us successive readings that differ systematically is not like a random error: the drift pushes the overall mean increasingly one way.

A similar argument implies that *any* pattern in our readings, not necessarily one manifested as a steady drift, should make us wary of claiming a reduction by $\sqrt{n}$ from the standard uncertainty of each value to the standard uncertainty of their mean.

Correlation between values in a sequence is measured by a number called the *autocorrelation*[14] and denoted by R. Unlike ordinary correlation, autocorrelation is a function of the separation of terms in a sequence of values. The terms in the sequence are assumed to have been obtained at equal intervals. We may call $R(1)$ the autocorrelation between the populations represented by the following two columns which have been derived from a *single* sequence of values:

'x'	'y'
first term	second term
second term	third term
third term	fourth term
. . .	. . .

---

[12] We have encountered ESDM previously in the notation $s_{\bar{x}} = s/\sqrt{n}$ (equation (5.56)) or $u(\bar{x}) = s/\sqrt{n}$ (equation (4.3)). The ESDM is also referred to in many statistical texts as the *standard error* of the mean.

[13] See the last paragraph in section 4.1.2.

[14] A sequence with significant autocorrelation is sometimes described as *serially* correlated or having serial correlation.

Similarly, $R(2)$ is the autocorrelation between the following two populations:

'x'	'y'
first term	third term
second term	fourth term
third term	fifth term
...	...

and so on for $R(3)$, $R(4)$, etc. This two-column arrangement of terms was shown in section 7.2.1. In general, $R(k)$, for $k > 0$, is the autocorrelation between terms separated by $k - 1$ intervening terms.

Note that $R(0)$ is always $+1$, since each row consists of identical values, thus

'x'	'y'
first term	first term
second term	second term
third term	third term
...	...

If the terms in a sequence fluctuate in a manner known as 'white noise',[15] the autocorrelation is zero (or close to zero) for all $R(k)$ where $k > 0$.

Figure 7.2(a) shows a white-noise sequence of 1000 values. They were drawn from a population with mean 0 and standard deviation 1. Figure 7.2(b) is a graph of the autocorrelation for this sequence: it is 1 for zero time-separation ($R(0)$ in the notation above), but immediately reduces to negligible values for non-zero time-separation. Sequences with this property do obey the $u(x)/\sqrt{n}$ rule for the ESDM, and the ESDM for a large number of readings can accordingly be negligibly small.

Some time sequences of values do not contain white noise and have significant autocorrelation between widely separated terms. Figure 7.3(a) shows a sequence of 170 readings of air temperature, taken once every 15 seconds, in one location in a temperature-controlled laboratory where the air temperature is permanently maintained at a nominal 20 °C. The readings were obtained using a platinum resistance thermometer, the temperature being indicated indirectly through the measurement of the temperature-sensitive resistance of a coil of platinum wire. The temperature was read to a precision of tenths of a millidegree. Although the air temperature was controlled, nevertheless, over 40 or so minutes, drift and oscillation over a range of slightly less than 0.2 °C were observed. The mean temperature was 20.065 °C and the standard deviation was 0.051 °C. In this case the ESDM is *not* $(0.051/\sqrt{170})$ °C.

---

[15] The name 'white noise' indicates that the spectrum of frequencies making up the noise is extremely broad; this is analogous to the colour 'white', which is composed of all the colours of the visible spectrum.

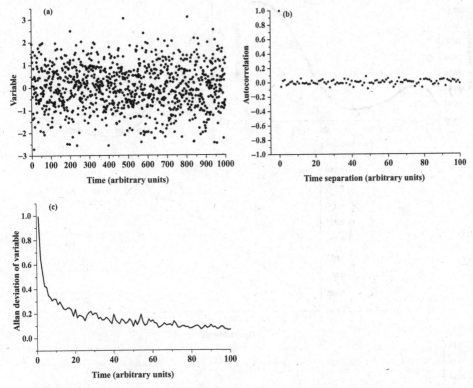

Figure 7.2. (a) 1000 uncorrelated readings from a Gaussian population: mean 0, standard deviation 1. (b) Autocorrelation of readings in (a). (c) The Allan deviation of readings in (a).

The reason can be seen when we plot the corresponding autocorrelation curve; it is shown in figure 7.3(b). Autocorrelation plots often follow this oscillation pattern from high positive to zero and then small negative values, followed by a slow return to zero. Here autocorrelation is significant (about +0.3 or higher) for time-separations up to about 9 minutes. If our readings had been taken at intervals of 15 minutes rather than 15 seconds, and n such readings had been collected, then the ESDM would have been reliably less than the standard deviation by a factor of $\sqrt{n}$. It is assumed that the temperature-control would have continued to operate over this much longer period.

In calculating autocorrelations by taking the 'x' and 'y' values from a single sequence, we have assumed that the sequence has the so-called 'ergodic' property (Bendat and Piersol 2000). The ergodic property implies, in general, that, if not just one but an *ensemble* of similar sequences is available for the same measurement procedure and under the same conditions, then mean values and autocorrelations over the entire ensemble at a particular time equal mean values and autocorrelations

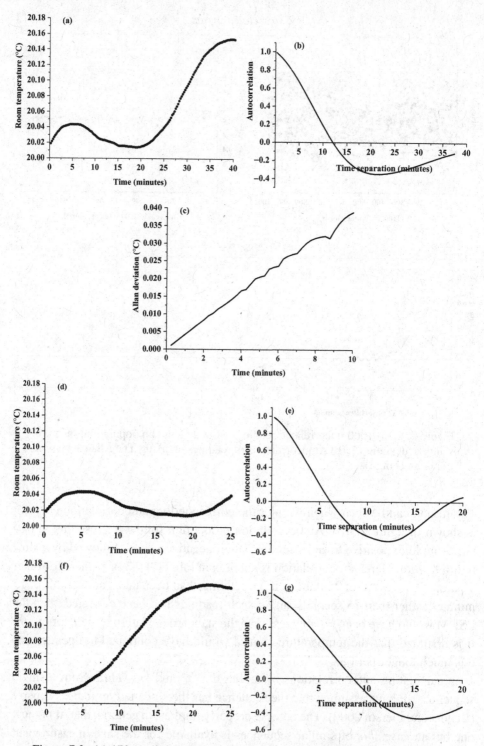

Figure 7.3. (a) 170 readings of air temperature taken every 15 seconds. (b) Auto-correlation of readings in (a). (c) The Allan deviation of readings in (a). (d) The first 100 points from (a). (e) Autocorrelation for the first 100 points. (f) The last 100 points from (a). (g) Autocorrelation for the last 100 points.

over one sequence over all times. For example, by calculating the autocorrelation, say $R(4)$ of *one* sequence between terms that are separated by three intervening terms (between first and fifth, second and sixth, etc.), the assumed ergodic property says that, if we were able to amass very many similar sequences (under the same conditions) and calculated the correlation of *only* the second and sixth terms (say) in each one, we would obtain the same result. Also the mean of one actual sequence, over all times, would be equal to the mean over all the possible sequences at a particular instant of time. The ergodic property says essentially that our *single* obtained sequence is faithfully *representative* of all the sequences we *might* have obtained.

We note that a sequence that presents a steady drift is not ergodic with respect to its mean value, since this obviously changes from one sequence to the next. However, the sequence is ergodic with respect to autocorrelations, and, in view of the perfect positive correlation for the case of a steady drift, equation (7.40) or $u(y) = u(x)$ holds for such a sequence.

Ergodic sequences belong to the class of *stationary* sequences, which can be described, roughly, as those sequences whose mean and autocorrelation do not depend strongly on our choice of starting or finishing points. The sequence of temperature measurements in the temperature-controlled laboratory shown in figure 7.3(a) is only roughly stationary. Thus, if we take only the first 100 points in figure 7.3(a), we have the graph of figure 7.3(d) with its autocorrelation shown in figure 7.3(e). If we take only the last 100 points in figure 7.3(a), we have the graph of figure 7.3(f) with its autocorrelation shown in figure 7.3(g). In the former case the autocorrelation remains significant for about 4 minutes, whereas in the latter the corresponding time is about 6 minutes.

Another way to characterise a sequence of values is by calculating the so-called 'Allan variance' and its square root, the Allan deviation (Allan 1987) (alternative names are the two-sample variance and two-sample standard deviation). In this procedure, we essentially gather together a group of *successive* readings in a sequence (the individual readings being separated by equal intervals), calculate their mean, and compare this mean with the mean of the next adjacent group of the *same* length. For this comparison, the squared difference of the means is calculated. The sum of all such squared differences between adjacent groups in the sequence, divided by twice the number of all such groups, is the Allan variance.

The Allan variance is, therefore, a function of the length of each group. If the sequence is a white-noise sequence, we expect the Allan variance to be inversely proportional to the length of each group. This is because, for uncorrelated readings as in white noise, the variance of their mean is inversely proportional to the length of the group (see, for example, equation (7.30)). Thus the longer the group, in a white-noise sequence, the smaller will be the (squared) differences

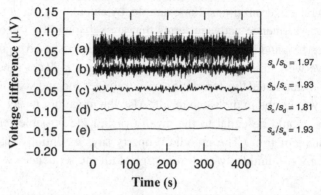

Figure 7.4. (a) A plot of 4096 successive voltage measurements made with an Agilent 34420A DMM with the input short-circuited. (b) A plot of the same data after grouping measurements into successive sets of four points and replacing the four points by the average value. Trace (c) is obtained by grouping the points in (b) into successive sets of four points and replacing the four points by the average value. Trace (d) is obtained by similar averaging of the points in (c) by sets of four. For white noise, we would expect that averaging by sets of four points would decrease the standard deviation of each plot with respect to that above it by a factor of two. The calculated ratios of successive standard deviations are given to the right of the plot. It can be seen that the ratios are slightly smaller than two (courtesy T. J. Witt, BIPM).

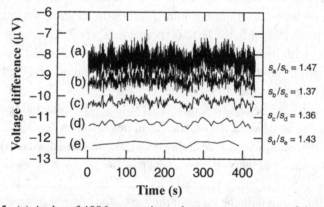

Figure 7.5. (a) A plot of 4096 successive voltage measurements of the difference between the 10-V outputs of two Zener-diode-based electronic voltage standards (Fluke 732B). The measurements were made with the same Agilent 34420A DMM as was used to gather the data appearing in figure 7.4. Trace (b) is a plot of the same data after grouping measurements into successive sets of four points and replacing the four points by the average value. Trace (c) is obtained by grouping the points in (b) into successive sets of four points and replacing these four points by the average value. Trace (d) is obtained by similarly averaging the points of (c) by sets of four. For white noise, we would expect that averaging by sets of four points would decrease the standard deviation of each plot with respect to that above it by a factor of two. In this case the noise is a mixture of $1/f$ noise and white noise and averaging by sets of four points reduces successive standard deviations by a factor of only about 1.4. The persistence of an irregular 'skeleton' of fluctuations is an indication of $1/f$ noise (courtesy T. J. Witt, BIPM).

between the means of such adjacent long groups. The Allan deviation of a white-noise sequence will, therefore, be inversely proportional to the square root of the length of the group. Figure 7.2(c) shows the Allan deviation as a function of the length (in this case, the length of time spanned by each group) for the same sequence of white-noise readings as in figure 7.2(a). Apart from the small fluctuations, the overall curve, in figure 7.2(c), has an inverse square-root dependence on time.

By contrast, figure 7.3(c) shows the Allan deviation for the highly correlated sequence of room-temperature readings of figure 7.3(a). There is a roughly linear increase in the Allan deviation, accompanied by oscillations of increasing amplitude.

In electronic circuits, white noise as in figure 7.2(a) is the natural variation in voltage across a resistance created by random thermal motion of electrons and known as 'Johnson noise'. Over a range of detected frequencies, or the 'passband', f_{pass}, the standard deviation, σ_J, of this noise in volts is $\sigma_J = \sqrt{4kTRf_{pass}}$, where k, T and R are the Boltzmann constant ($k \simeq 1.38 \times 10^{-23}$ J/K), absolute temperature and resistance, respectively. Thus for $R = 10\,000\ \Omega$ and $T = 293$ K (approximately room temperature) $\sigma_J \simeq 13$ nV over 1 Hz of bandwidth. We note that, if we, say, double the passband, we also double the variance, σ_J^2, of the detected noise. This is a characteristic of white noise.

Another type of noise is also common in electronic circuits. This is so-called $1/f$ noise, which, as the name implies, increases as the frequency is lowered and is roughly inversely proportional to it. A plot of voltage readings against time, for $1/f$ noise, shows autocorrelations and, once again, the ESDM cannot be obtained from the standard deviation by division by $\sqrt{n}$. This spectrum of noise is observed in voltage standards based on Zener diodes (Witt and Reymann 2000) and in superconducting devices known as SQUIDs (superconducting quantum interference detectors), which are used as sensitive detectors of tiny magnetic fields (Cantor and Koelle 2004). No increase in stability is observed when a group of individual readings is replaced by their mean, nor when such a process of averaging is repeated. The Allan deviation of $1/f$ noise when plotted against time is a horizontal line and so is somewhere intermediate between the cases illustrated in figures 7.2(c) and 7.3(c). Figures 7.4 and 7.5 show the effects of successive averaging applied to white noise and to a mixture of white and $1/f$ noise, respectively (Witt 2000).

When a sequence exhibits autocorrelations, a simple and safe option is to characterise the ESDM as equal to the standard deviation. There exists a range of more complicated procedures. Among the simplest of these is the use of 'binary grouping' or 'binary blocking' of a sequence of n readings where n is a power of 2 (Flyvbjerg and Petersen 1989).

7.2.3 Testing for autocorrelation in a short sequence of readings

Very often readings of the same quantity are obtained manually, rather than by means of automated instruments. Unless the experimenter has much time and patience, only a few values are obtained. We therefore consider the question of detecting the presence or absence of autocorrelation in a short sequence of n readings, and in particular whether dividing the standard deviation by $\sqrt{n}$, to obtain the ESDM, is justifiable.

The presence of any pattern in the readings, not necessarily a steady drift, may indicate autocorrelation. Such a pattern may be, for example, a steady drift, a quadratic (or higher-order) dependence on time, or part or whole of a sinusoid. With any pattern, the successive readings might not be independent; they may present a mutually 'sticky' quality, such that it becomes possible, having taken, say, ten or so successive readings, to discern a rough trend and so to predict with some accuracy where the next reading is likely to be in relation to them. Although a lack of independence does not imply the presence of correlation (whereas independence does imply zero correlation),[16] nevertheless, in most practical cases, if we observe that a reading depends to some extent on previous readings, we may assume that autocorrelation exists. It is usually not possible with short sequences to quantify this autocorrelation reliably. Moreover, manual readings are often obtained without particular regard for the need to have at least roughly equal intervals. A safe practice if correlation is suspected, which avoids the risk of an unrealistically small standard uncertainty in the mean, is to use equation (7.40), which implies taking the standard deviation of the readings as the ESDM.

Short sequences are often not pure time-sequences but may also be sequences in space or some other variable that is deliberately varied. In measuring the temperature coefficient of some physical property, for example (like length or electrical resistance), that property is measured several times at intentionally different temperatures. The profilometer readings in exercise D in section 7.1.2 involve a sequence not only in time but also in space. If, to take a hypothetical case, a profile forms a slope, the spatial analogue of a steady drift in time, it is plain that, just as for a drift with its high positive autocorrelation, the mean thickness of the slope can be assigned a standard uncertainty equal to the standard deviation of the thickness over the measured range, with no reduction by $\sqrt{n}$.

When a sequence reveals a pattern, we may choose to fit parameters to it by least-squares. When the pattern is a simple one such as a slope or smooth curve, the results of the fit are generally more informative than the standard deviation of the raw readings. The rate of drift, b, of a quantity can be estimated (see equation (5.53)), and any random fluctuations superimposed on the drift will contribute to

---

[16] An example of the difference between independence and zero correlation was given in section 5.3.

the standard uncertainty in b. Using $u(b)$ to represent the standard uncertainty in b, we have

$$u(b) = s\sqrt{\frac{n}{n\sum_{i=1}^{n} x_i^2 - \left(\sum_{i=1}^{n} x_i\right)^2}}, \tag{7.41}$$

where s is the root-sum-square residual and the x_i $(i = 1, 2, \ldots, n)$ are the assumed error-free predictor or explanatory variables. Equation (7.41) may also be written

$$u(b) = s\frac{\sqrt{n}}{n \times \text{standard deviation of } x}. \tag{7.42}$$

We know that s is relatively insensitive to the number, n, of readings.[17] For a given set of values of x (which is the explanatory variable whose error-free values we can choose), equation (7.42) therefore shows that $u(b)$ varies as $\sqrt{n}/n = 1/\sqrt{n}$, just like the ESDM of uncorrelated readings. Such a $1/\sqrt{n}$ dependence is a general characteristic of the standard uncertainty of least-squares estimates, of which the mean is the simplest example. Ideally, fitting parameters by least-squares should remove the autocorrelation that creates a pattern and should yield uncorrelated residuals, thereby restoring the reduction by $\sqrt{n}$ in going from the root-mean-square residual, s, to the standard uncertainty of the fitted parameters. If a pattern can still be discerned among the residuals to a least-squares fit, the particular least-squares model is inadequate; for example, a higher-order model may need to be considered rather than a linear fit.[18]

7.2.4 Reduction in uncertainty of measurand due to correlated inputs

Correlations between inputs can also work to our advantage in *reducing* the uncertainty in the measurand. Suppose that there are two inputs, x_1 and x_2, and that they are highly positively correlated. More precisely, as previously mentioned, this means that the *errors* in the inputs are highly positively correlated. We can then take $r(x_1, x_2) = +1$ to a good approximation. Let the measurand, y, be the difference between the two inputs:

$$y = x_1 - x_2. \tag{7.43}$$

Since $\partial y/\partial x_1 = 1$ and $\partial y/\partial x_2 = -1$, equation (7.37) gives

$$u^2(y) = u^2(x_1) + u^2(x_2) - 2u(x_1)u(x_2). \tag{7.44}$$

---

[17] For example, if we double the number of points on the graph, we do not expect to find twice the amount of scatter as before. The standard deviation of a set of readings has a similar property of low sensitivity to the number of readings (from the same population).

[18] Tests for autocorrelation are discussed in Draper and Smith (1981).

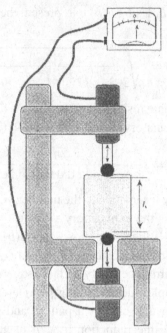

Figure 7.6. A gauge-block comparator (courtesy J. E. Decker and J. R. Pekelsky (1996), National Research Council of Canada).

Since the right-hand side is a perfect square,

$$u(y) = u(x_1) - u(x_2). \tag{7.45}$$

If, therefore, x_1 and x_2 are measured using the same instrument, and are of similar magnitude, so that $u(x_1)$ and $u(x_2)$ are likely to be approximately equal, equation (7.45) implies that

$$u(y) \sim 0. \tag{7.46}$$

Examples of uncertainty-reducing high correlation are quite common. If a person monitors his or her weight on the same set of bathroom scales, and x_1 and x_2 are the weights at two different times, then the fact that the scales may have a systematic error is scarcely important: they will correctly register any loss or gain in weight between these two times. We observe here another interpretation of a systematic error: it may be regarded as a random error with a much longer time-constant than the repetition interval of measurements.

The low uncertainty offered by difference measurements between highly positively correlated inputs is exploited in many fields of metrology. Figure 7.6 shows a schematic diagram of a gauge-block comparator as used in length metrology. The measured length is that recorded between the opposing styli, which penetrate to a

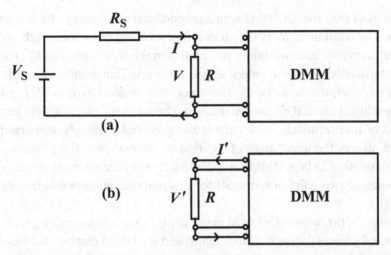

Figure 7.7. (a) Measurement of V by DMM; (b) measurement of R by DMM.

small extent (a few tens of nanometres) into the material of the gauge block (often tungsten carbide or steel). This penetration affects the accuracy of the measurement of the thickness of the gauge block. However, the comparison of different gauge blocks, of the same material and therefore undergoing similar amounts of stylus penetration, is relatively insensitive to the penetration depth. For similar reasons, such a comparison is relatively insensitive to small changes in ambient temperature arising during the comparison.

Suppose that we wish to measure with high accuracy a current, I, passing through a resistance, R. To do this we measure the voltage, V, across the resistor and use Ohm's Law: $I = V/R$. Here I is the measurand, and V and R are the input quantities. Uncertainties in V and in R will propagate into I, creating an uncertainty in I. We have $\partial I/\partial V = 1/R$ and $\partial I/\partial R = -V/R^2$, so that, if V and R are uncorrelated, we may use equation (7.14) to obtain the standard uncertainty, $u(I)$, of the current in terms of the standard uncertainties, $u(V)$ and $u(R)$, in V and R, respectively. Equation (7.14) then gives

$$u^2(I) = \frac{1}{R^2}u^2(V) + \frac{V^2}{R^4}u^2(R). \tag{7.47}$$

However, we need to discuss whether there is likely to be any correlation between V and R. We assume that the electric circuit for measuring V is as shown in figure 7.7(a), and that the circuit for measuring R is as shown in figure 7.7(b). The instrument is a digital multimeter or DMM that can measure resistance and current as well as voltage. In this application, the DMM is required to measure voltage and resistance. High-quality DMMs can measure voltages of the order of 1 V and resistances of the order of 100 Ω with a proportional uncertainty of a few parts per million. The resistance, R, is shown as a four-terminal resistance, with two outer 'current' terminals and two inner 'potential' terminals. If a current I is fed to the current terminals, so that I enters at one current terminal and exits at the other current terminal, the value of the resistance R is *defined* as $R = V/I$, where V is the resultant potential difference measured between the two potential terminals. The use of four terminals, with current and potential terminals deliberately kept separate, avoids the uncertainty of location of the two potential points in a two-terminal resistor.[19] Many DMMs are able to measure four-terminal resistances and have therefore two pairs of terminals for this purpose, as shown in figures 7.7(a) and 7.7(b).

In figure 7.7(a), where the DMM measures V, a voltage source, V_S, with output resistance R_S, passes current, I, through R, and the DMM displays the value of V. Only one of the two pairs of DMM terminals is needed for this measurement. In figure 7.7(b), the DMM measures R. To do so, the other pair of DMM terminals provides a standard current, I', through R, whereupon the DMM measures the resultant V' and displays (using an internal algorithm) the value of R as $R = V'/I'$. The required value of the measurand I is then given by $I = V/R$.

Suppose that the standard current, I', is roughly equal to I. Then V and V' will also be roughly equal. If the same DMM is used in figures 7.7(a) and 7.7(b), the errors δV and $\delta V'$ are therefore likely to be of the same sign and roughly equal. In figure 7.7(b), the displayed value of R is given by $R = V'/I'$, so the error, δR, in R is given by $\delta R = \delta V'/I' \sim \delta V/I'$. In practice there will be an additional uncertainty in the standard current I', but this argument shows that, if the same DMM is used in figures 7.7(a) and 7.7(b), then δR and δV are likely to be highly positively correlated. If the correlation coefficient $r(V, R) \sim +1$, equation (7.37) gives

$$u^2(I) = \frac{1}{R^2}u^2(V) + \frac{V^2}{R^4}u^2(R) - \frac{2V}{R^3}u(V)u(R) \qquad (7.48)$$

and the right-hand side is now a perfect square, so that equation (7.48) gives

$$u(I) = \frac{1}{R}u(V) - \frac{V}{R^2}u(R). \qquad (7.49)$$

So, by using the same DMM in figures 7.7(a) and 7.7(b), we can, in principle, achieve

$$u(I) \sim 0 \qquad (7.50)$$

---

[19] In electrical metrology, four-terminal connections are needed when high accuracy is required, such as in the case of the 1-ohm standard resistor in figure 3.2.

so long as $u(V)/V = u(R)/R$, so that the proportional uncertainty in the displayed voltage V (in figure 7.7(a)) equals the proportional uncertainty in the displayed resistance R (in figure 7.7(b)). In this electrical example, the advantage afforded by the high positive correlation lies in the fact that the error in a ratio cancels out to zero if both the numerator and the denominator of that ratio have the same proportional error.[20]

We see that a positive correlation between two inputs to a measurand generally arises when the same instrument is used in measuring the values of both inputs. An additional condition (not always necessary) for a positive correlation is that the inputs have very roughly comparable values (say to within an order of magnitude). The instrument is then likely to be used in the same measuring range for both measurements, and consequently any systematic error in the instrument is likely to have the same value for both measurements.

7.3 Review

In this chapter we have considered how uncertainties propagate in situations where the errors in input quantities are uncorrelated as well as when errors are correlated. Irrespective of whether uncertainties are evaluated through statistical analysis (and hence are Type A uncertainties) or have been evaluated by other means (and are therefore Type B uncertainties), the method for combining them makes no distinction between types. In the next chapter we consider the probability of a particular value occurring when we make a measurement and how, in many cases, the distribution of values obtained in an experiment can be well described by a very important theoretical distribution, known as the 'Gaussian' or 'normal' distribution.

---

[20] This statement would not be correct if the word 'error' were replaced by 'uncertainty'. It is the errors, *not* the uncertainties, that are highly positively correlated. We see again the usefulness of the distinction between 'error' and 'uncertainty'.

8

Probability density, the Gaussian distribution and the central limit theorem

After measurement, we assign an estimated value to a measurand as well as an accompanying uncertainty. The uncertainty is usually expressed as an *interval* around the estimated value. With any such interval we associate a probability that the actual or true value of the measurand falls within that interval.[1] Measurands are usually *continuous* quantities such as temperature, voltage and time. However, when discussing probabilities in the context of measurement it is convenient first to consider 'experiments' in which the outcomes are *discrete*, for example tossing a coin, where the outcome is a head or a tail.

8.1 Distribution of scores when tossing coins or dice

A fair coin falls heads up with probability $\frac{1}{2}$ and tails up also with probability $\frac{1}{2}$. A fair coin is an idealised object (since all real coins have a slight bias towards either heads or tails) and presents the simplest case of a 'uniform' probability distribution. When a probability distribution is uniform, the possible outcomes of an experiment (tossing a coin in this case) occur with equal probability. We will show how non-uniform probabilities emerge as soon as two or more fair coins are considered. These non-uniformities tend to a characteristic pattern called a Gaussian (or 'normal') probability density distribution.[2] For the sake of brevity we shall usually refer to the 'Gaussian probability distribution' as simply the 'Gaussian distribution'. Likewise we shall usually refer to the 'uniform probability density distribution' as the 'uniform distribution'.

Given a coin, it is convenient to assign a score to the result of each toss: $+1$ for heads and -1 for tails. If only one coin is tossed, the possible scores will be $+1$,

---

[1] Thus if the measurand is the diameter of a metal rod and is estimated to be 25.37 mm with an uncertainty quoted as ± 0.06 mm, we infer that there is a high probability, commonly 95%, that the diameter lies in the interval 25.31 mm to 25.43 mm. We shall see in chapter 10 that an uncertainty expressed in this way, with a $\pm$ sign, is a so-called *expanded* uncertainty.

[2] Named after Karl-Friedrich Gauss (1777–1855).

with probability $\frac{1}{2}$, and -1, also with probability $\frac{1}{2}$. These probabilities[3] sum to 1, meaning that it is certain that we shall get one or other of these mutually exclusive scores.[4]

If two coins are tossed, the outcomes and scores are (where H represents a head and T a tail)

HH	+2
HT	0
TH	0
TT	−2

Of the four possible outcomes ($2^2 = 4$), a score of zero appears twice and so has probability $\frac{2}{4} = \frac{1}{2}$. The score of +2 appears only once and therefore has a probability of $\frac{1}{4}$. Similarly for the score of −2. The sum of the three probabilities is $\frac{1}{2} + \frac{1}{4} + \frac{1}{4} = 1$. Again, it is certain that we shall obtain one of these mutually exclusive scores.

If three coins are thrown, the outcomes and scores are

HHH	+3
HHT	+1
HTH	+1
HTT	−1
THH	+1
THT	−1
TTH	−1
TTT	−3

Out of eight outcomes ($2^3 = 8$), the score of +1 appears three times and so has probability $\frac{3}{8}$. Similarly for a score of −1. The less likely scores of +3 and −3 each have a probability of $\frac{1}{8}$. The sum of the four probabilities is $\frac{3}{8} + \frac{3}{8} + \frac{1}{8} + \frac{1}{8} = 1$.

It is straightforward, if rather tedious, to go through a similar procedure for finding the possible scores and their probabilities for four or more coins. With n coins, there are 2^n outcomes. If there are h heads in any one of these, the score, S, for that outcome is

$$S = 2h - n, \tag{8.1}$$

[3] The probability, P, of an event is always a positive number between 0 and 1. The larger P, the more probable the event. $P = 0$ for an impossible event, and $P = 1$ for an event that is certain. P is often expressed as a percentage, thus $P = 0.95$ (a highly probable event) may be written as $P = 95\%$.

[4] Since it is not possible to have as an outcome *both* a head *and* a tail on a single toss of a coin, these outcomes are said to be mutually exclusive. (We ignore the very small probability that the coin might land and balance on its edge!)

and the probability, $P(S)$, of that score is

$$P(S) = \frac{1}{2^n} \frac{n!}{h!(n-h)!}. \tag{8.2}$$

The symbol ! represents the factorial of a positive integer: the product of that integer and all smaller integers down to 1. Thus, for an integer m, $m! = m \times (m-1) \times (m-2) \cdots \times 2 \times 1$. For example, $5! = 5 \times 4 \times 3 \times 2 \times 1 = 120$. The expression $P(S)$ is a particular case of the binomial distribution.[5]

The situation is depicted in figure 8.1 for 1, 2, 3, 5, 8 and 20 coins. The 'envelope' of the array of probabilities approaches more and more closely the typical 'bell-shape', otherwise known as the 'Gaussian' or 'normal' shape, as the number of coins is increased. This shape does not depend on our arbitrary choice of scores of $+1$ for heads and -1 for tails; any other choice shifts the whole shape left or right (so that its peak would no longer be at zero), and may change its scale (width and height). However, the essential 'bell-shape' would remain. This general shape is shown in figure 8.5.

If, instead of coins, we have six-sided fair dice, the probability distribution gives a faster approach to the Gaussian shape as the number of dice increases. This is illustrated in figure 8.2 for throws of 1, 2, 3 or 4 dice, where scores are calculated in the conventional way as the sum of the number of dots on the uppermost faces. As players of dice-based board games know, the score of 7 is the most common score when two dice are used, because 7 can be obtained in more ways than any other score $(6+1, 1+6, 5+2, 2+5, 4+3, 3+4)$. So 7 is the peak value in figure 8.2(b), occurring with a probability $\frac{6}{36} = \frac{1}{6}$. (The total number of outcomes with two six-sided dice is $6^2 = 36$.) Just as in figure 8.1, the sum of the probabilities in each of figures 8.2(a)–(d) is 1.

Exercise A

If ten fair coins are tossed, what are the probabilities of obtaining
   (a) five heads and (b) fewer than three heads?

8.2 General properties of probability density

In the examples of the coins and dice, the score varies in discrete steps, and so does the probability. However, most physical quantities vary continuously. In these cases we need to consider a *probability density* rather than a probability. We have

---

[5] The name 'binomial' expresses the fact that there are only two possible outcomes of a trial (in our example the outcome is a head or tail) for each of n trials (the toss of a coin is regarded as a trial). The general binomial case involves different probabilities p for success and $1-p$ for failure; in our examples, $p = \frac{1}{2}$ for a fair coin.

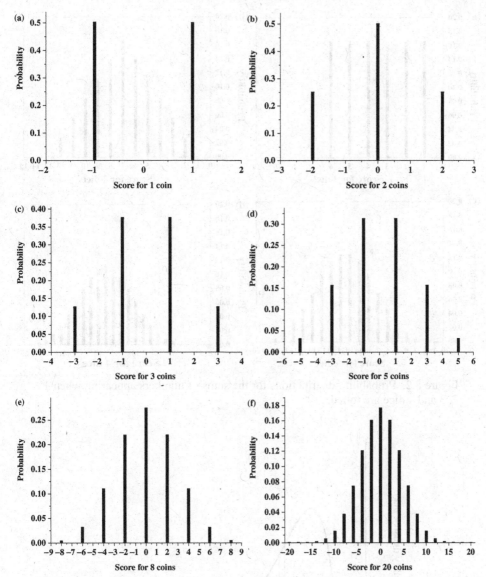

Figure 8.1. Probability distributions of the scores obtained by tossing 1, 2, 3, 5, 8 and 20 coins.

previously denoted probability by an upper-case P; probability density will be denoted by a lower-case p.

Figure 8.3 shows a possible form of a graph of the probability density, $p(x)$, of the continuous random variable x. The graph describes the *probability density distribution* of x, or probability density function (pdf) of x. Briefer names are the distribution or density distribution of x. The probability that x lies in the interval x

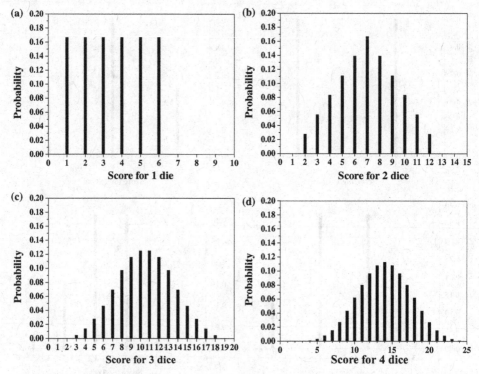

Figure 8.2. Probability distributions for the sums of numbers appearing when 1, 2, 3 and 4 dice are rolled.

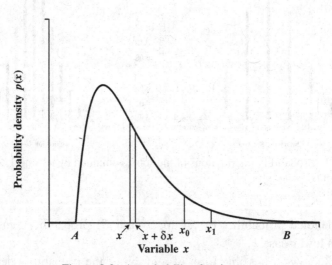

Figure 8.3. A probability density curve.

to $x + \delta x$ is equal to the area of the narrow vertical strip under the curve in figure 8.3 between x and $x + \delta x$. This area is[6] $p(x)\delta x$. The probability that x takes a value between more widely separated points such as x_0 and x_1 is the area expressed as the integral, $\int_{x_0}^{x_1} p(x)\,dx$. Since $p(x)$ is largest at the peak of the probability density curve, the probability of obtaining a value in a given interval of x is greater the closer that interval is to the peak. By contrast, in a region where $p(x) = 0$, for example at $x < A$, the probability of obtaining a value of x in that region is zero.[7]

Since $p(x)$ is a probability density, the product of $p(x)$ and a range of x is a probability: it is a dimensionless number between 0 and 1. It follows that the dimensions of a probability density $p(x)$ are the inverse of the dimensions of x.[8]

A probability density generally describes a population rather than a sample. Important attributes of any population are its mean and standard deviation. We have encountered several alternative but equivalent expressions for each of these. For example, μ, μ_x and $E(x)$ have each been used to represent the population mean of x. We now introduce another representation of the mean in terms of the probability density, $p(x)$.

We first note that $\int p(x)\,dx = 1$, where the integral is over the entire permitted range of x (where $p(x) \neq 0$). In figure 8.3, this is the range $x = A$ to $x = \infty$. There are cases, as in the Gaussian probability density distribution, where x can vary anywhere between minus infinity and plus infinity; we then have

$$\int_{-\infty}^{+\infty} p(x)\,dx = 1. \tag{8.3}$$

Equation (8.3) can be taken to include the case of a finite permitted range, as in figure 8.3, provided that $p(x)$ is set equal to zero outside this permitted range. Equation (8.3) then states that it is certain (the probability is equal to 1) that x must lie somewhere within its permitted range. Equation (8.3) states, equivalently, that the total area underneath the probability density curve must be 1.

The mean, μ, can now be written as

$$\mu = E(x) = \int_{-\infty}^{+\infty} x p(x)\,dx. \tag{8.4}$$

Equation (8.4) states that the mean of x is the sum of the possible values of x, each weighted by the probability that x takes that value. The following example in terms

---

[6] This assumes that the strip is rectangular, of height $p(x)$ and width δx. In fact the strip is not rectangular, since the lower edge is horizontal but the upper edge has a slope. However, the error involved is only second order (involving $(\delta x)^2$), and is negligible.

[7] When x is a continuous variable, it is worth noting that the probability that x should take a *particular* value, having in effect a zero associated interval, is zero; only *intervals* of x, whether small or large, can have non-zero probabilities. The relationship between probability density and probability is analogous to that between ordinary density and mass.

[8] For example, if x represents a length, then the dimensions of the probability density would be (length)$^{-1}$.

of discrete probabilities (and a very small population) illustrates the soundness of this method of determining the mean.

Suppose that a population consists of seven discrete values, 1, 1, 1, 1, 2, 2, 3. The mean of these values is $\mu = \frac{11}{7}$. The probability, $P(1)$, of choosing the value 1 in the population is $P(1) = \frac{4}{7}$. Similarly, $P(2) = \frac{2}{7}$ and $P(3) = \frac{1}{7}$. For this discrete case, analogously to equation (8.4), we have

$$\mu = E(x) = 1 \times P(1) + 2 \times P(2) + 3 \times P(3)$$
$$= 1 \times \frac{4}{7} + 2 \times \frac{2}{7} + 3 \times \frac{1}{7} = \frac{11}{7}.$$

Equation (5.11) in chapter 5 expresses the variance, σ^2, of a population as the mean square minus the squared mean, so we may write

$$\sigma^2 = \int_{-\infty}^{+\infty} x^2 p(x)\, dx - \left(\int_{-\infty}^{+\infty} x p(x)\, dx \right)^2, \tag{8.5}$$

and the standard deviation of the population is the square root of equation (8.5). The first term on the right-hand side of equation (8.5) is $E(x^2)$, the mean value of x-squared (analogous to equation (8.4) for the mean of x):

$$E(x^2) = \int_{-\infty}^{+\infty} x^2 p(x)\, dx. \tag{8.6}$$

The counterpart to equation (8.6) in our discrete example above is

$$E(x^2) = 1^2 \times P(1) + 2^2 \times P(2) + 3^2 \times P(3)$$
$$= 1 \times \frac{4}{7} + 4 \times \frac{2}{7} + 9 \times \frac{1}{7} = \frac{21}{7} = 3.$$

This may be verified by squaring each of the seven values and taking the mean of these squares. We finally have that $\sigma^2 = E(x^2) - (E(x))^2 = 3 - \left(\frac{11}{7}\right)^2 = \frac{26}{49}$, or $\sigma = \frac{\sqrt{26}}{7} \simeq 0.73$.

Equation (5.5) in chapter 5, repeated here, may be shown to give the same result:

$$\sigma^2 = \frac{\sum_{i=1}^{N}(x_i - \mu)^2}{N},$$

and, with $N = 7$ in our example, we have

$$\sigma^2 = \frac{1}{7}\left[\left(1 - \frac{11}{7}\right)^2 + \left(1 - \frac{11}{7}\right)^2 + \left(1 - \frac{11}{7}\right)^2 + \left(1 - \frac{11}{7}\right)^2 \right.$$
$$\left. + \left(2 - \frac{11}{7}\right)^2 + \left(2 - \frac{11}{7}\right)^2 + \left(3 - \frac{11}{7}\right)^2 \right]$$
$$= \frac{1}{7}\left(4 \times \frac{16}{49} + 2 \times \frac{9}{49} + \frac{100}{49} \right) = \frac{1}{7} \times \frac{182}{49} = \frac{26}{49},$$

agreeing with σ^2 obtained previously.

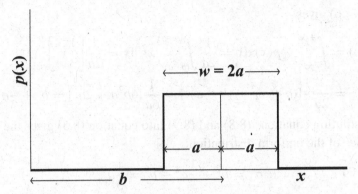

Figure 8.4. A uniform or rectangular probability distribution.

Exercise B

(1) A population consists of ten discrete values: 3, 3, 5, 5, 5, 6, 7, 8, 8, 8. Find the mean, standard deviation and variance of these values.

(2) A particular probability density can be written $p(x) = Ax$ for the range $0 < x < 2$ and $p(x) = 0$ outside this range.
(a) Sketch the graph of $p(x)$ versus x.
(b) Determine the constant, A.
(c) Calculate the probability that x lies between $x = 1$ and $x = 1.5$.

8.3 The uniform or rectangular distribution

The simplest example of a probability density is the so-called uniform or rectangular probability density. In this case, the probability density is zero everywhere except in a particular region, and in this region $p(x)$ is a positive constant. Figure 8.4 illustrates the case where $p(x)$ is centred on $x = b$ and has a constant value from $x = b - a$ to $x = b + a$. The shape of the distribution is rectangular, hence one of its names.

The 'height' of the distribution in figure 8.4 must be $1/(2a)$. This follows from the condition expressed by equation (8.3) that the area enclosed by the rectangle must be 1, and from the horizontal extent, $2a$, of the rectangle. Thus the uniform distribution is described as

$$p(x) = \begin{cases} 1/(2a), & b - a < x < b + a, \\ 0, & \text{for all other values of } x. \end{cases} \tag{8.7}$$

The symmetry of the distribution in figure 8.4 indicates that the mean, μ, is given by $\mu = b$. This can be shown more formally using equation (8.4) as follows:

$$\mu = \int_{-\infty}^{+\infty} xp(x)\, dx = \frac{1}{2a} \int_{(b-a)}^{(b+a)} x\, dx = \frac{1}{2a} \left[\frac{1}{2} x^2 \right]_{(b-a)}^{(b+a)}$$

$$= \frac{1}{2a}\frac{1}{2}[(b+a)^2 - (b-a)^2] = \frac{1}{4a}(4ba) = b. \tag{8.8}$$

Equation (8.6) gives

$$E(x^2) = \int_{-\infty}^{+\infty} x^2 p(x)\, dx = \frac{1}{2a} \int_{(b-a)}^{(b+a)} x^2\, dx = \frac{1}{2a}\left[\frac{1}{3}x^3\right]_{(b-a)}^{(b+a)}$$

$$= \frac{1}{2a}\frac{1}{3}[(b+a)^3 - (b-a)^3] = \frac{1}{6a}[6b^2a + 2a^3] = b^2 + \frac{1}{3}a^2. \quad (8.9)$$

Thus substituting equations (8.8) and (8.9) into equation (8.5) gives the result for the variance of the uniform distribution:

$$\sigma^2 = b^2 + \frac{1}{3}a^2 - b^2 = \frac{1}{3}a^2, \qquad (8.10)$$

or for its standard deviation:

$$\sigma = a/\sqrt{3}. \qquad (8.11)$$

A uniform distribution of 'half-width', a, therefore has a standard uncertainty $u = a/\sqrt{3}$ (recalling that standard deviation and standard uncertainty are equivalent). Sometimes the full-width, $w = 2a$, is more convenient, in which case the standard uncertainty is expressed as $u = w/\sqrt{12}$. The standard uncertainty is independent of the location, b, of the centre of the uniform distribution. In many cases the uniform distribution is centred on zero, so that $b = 0$.

A uniform distribution in metrology arises more often as an expression of our ignorance, rather than as a description of observable fact. A case in point arises when a continuous variable, such as a voltage, is measured and displayed by a digital multimeter (DMM). Suppose that the DMM displays only four decimal digits and that the display is 3.571 V. Then the actual reading may be anywhere, and with uniform probability, within the (approximate) interval 3.5705 V to 3.5715 V. We accordingly have $w = 0.001$ V, or $a = 0.0005$ V. The standard uncertainty arising from limited resolution is given by $a/\sqrt{3} \simeq 0.000\,29$ V or about 290 μV. In general, when *all* we know about a quantity are its lower and upper bounds – as in the case of a limited-resolution digital display – a uniform distribution between these two bounds can legitimately be assumed and has theoretical backing.[9]

The distribution of the errors that make up a Type B uncertainty is sometimes claimed to be uniform. The supporting argument is that, there being no statistical treatment available such as would be provided by usefully repeated measurements, all that is known are the end-points within which the quantity can plausibly vary; hence it must be uniformly distributed between them. This argument is flawed when the value of the quantity and its uncertainty are the subject of a calibration report or

---

[9] Another case where the uniform distribution is generally assumed to be applicable is in microwave metrology, when at high frequencies the phase shift of a reflected signal is unknown except for being limited to the range $0°$ to $360°$. Further discussion on the occurrence of the uniform distribution in metrology may be found in Cox and Harris (2004).

Table 8.1. *Resolutions of several instruments*

Instrument	Resolution
Thermometer	0.5 °C
Measuring cylinder	0.2 mL
Capacitance meter	10 pF
Stopwatch	0.01 s

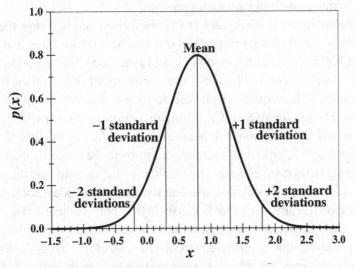

Figure 8.5. Gaussian probability density with mean $\mu = 0.8$, standard deviation $\sigma = 0.5$.

have been determined from a look-up table; in such a case the quantity will have the distribution observed or postulated by the compiler of the report or look-up table, and this is likely to be Gaussian, or approximately so.

Exercise C

Table 8.1 includes several instruments together with their limits of resolution. The 'limit of resolution' was represented by the symbol w above. For each instrument calculate the standard uncertainty due to the limit of resolution to two significant figures.

8.4 The Gaussian distribution

8.4.1 Gaussian distribution of measurement errors

The most important and commonly observed distribution is the Gaussian. The probability density distribution is shown in figure 8.5 and is recognisable as the envelope of the discrete probabilities for scores obtained with a large number of coins or dice shown in figures 8.1 and 8.2. The particular case of a Gaussian shown in figure 8.5 has a mean $\mu = 0.8$ and standard deviation $\sigma = 0.5$.

The essential physical process that in metrology creates a Gaussian distribution of errors can be discerned from the examples of the coins and dice in section 8.1. What we called the 'score' in these examples corresponds to the error in a measurement. The score is the arithmetical sum of more elementary constituents, such as the face-up value of *one* particular coin among several tossed coins. The error in a measurement is, similarly, the sum of many independent but simultaneously acting random contributions from various sources.

In the case of a measurement, each error contribution may lie below the threshold of observation. For the total error to be large and positive (or large and negative), these contributions must act, fortuitously, all in the same direction. This will happen rarely, since the contributions act independently of one another. In this way we can explain, at least qualitatively, the thinly populated 'tails' of the Gaussian distribution. Thus in figure 8.1(f), referring to a throw of 20 coins, a score of $+20$ can happen only if all 20 coins fall heads; the probability of this is $1/2^{20} \sim 10^{-6}$. Similarly, in figure 8.2(d) when four dice are thrown, the outcome may be a score of 4, but for this to happen all four dice must fall with 1 face-up, and the probability of this is $1/6^4 < 10^{-3}$. By contrast, the simultaneous independent contributions are much more likely, at any given moment, to comprise both positive and negative contributions in roughly equal numbers, creating a small net error. We therefore have a qualitative explanation for the well-populated peak of the Gaussian distribution.

Intuitively, we may regard a Gaussian distribution as the natural distribution of the observable combined outcome of additive, independently acting and not directly observable influences of randomly varying sign. This is why the errors in a measurement are often assumed by default to have a Gaussian distribution. It is common to find experimentally that random errors, measured as the differences between measured values and their mean, or more generally as residuals from a least-squares fit, have the following properties:

(i) large values of random error, whether positive or negative, occur less frequently than small values; and

(ii) positive and negative values of random error occur more or less equally often and are, roughly, symmetrically disposed around zero.

Such a distribution has an approximate 'bell-shape', peaked at zero, and is generally considered to be an approximate real-world representation of the Gaussian distribution.

A Gaussian distribution does not necessarily describe errors, in the metrological sense of an unwanted presence that should be avoided or reduced as much as

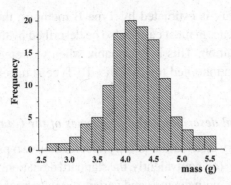

Figure 8.6. Cocos-palm fruit mass: mean 4.20 g, standard deviation 0.50 g.

possible. It may also describe the natural distribution of some attribute of a population (where 'population' may have its everyday meaning). The height of adult humans of each sex and ethnic group follows an approximate Gaussian distribution, governed by many influences that may be grouped broadly as genetic and environmental. In chapter 5 we considered a sample of six pieces of fruit from a palm-tree. In fact, 120 pieces were calculated and weighed; the distribution is shown as a histogram in figure 8.6, which approximates a Gaussian shape.

The strong theoretical underpinning of the Gaussian distribution – briefly stated, as the natural additive combination of small random influences – together with this common experimental finding, explain the frequently used alternative term 'normal' distribution. We shall occasionally use the term 'normality' to refer to the Gaussian property of a distribution.

In the example of the measurement of the temperature coefficient of resistance of a standard resistor (figure 4.1), the scatter of the values at a given temperature can be explained partly as the effect of electronic noise and electromagnetic interference on the digital multimeter (DMM) used to measure the resistance (by comparison with another standard resistor at a fixed temperature). This noise and interference affect the display of the DMM and may be regarded as contributing small random voltages to the DMM. Such contributions would, again, be relatively unlikely mutually to reinforce one another, and more likely partially to cancel each other.

The errors referred to above, and likened to the total score in throwing coins or dice, are regarded as random errors. Section 4.2 defined 'uncertainty' as a measure of dispersion of values, and in section 4.3 the standard deviation was recruited as a measure of uncertainty and given the name 'standard uncertainty'. It is now clear that the standard deviation of the Gaussian distribution, depicted as an envelope to the probabilities in figures 8.1 and 8.2, is a natural measure of the Type A standard uncertainty created by random errors.

When an uncertainty is estimated by Type B methods, the associated Type B standard uncertainty can, in most cases, also be described by the standard deviation of a Gaussian distribution. This is reasonable when we remember that a Type B uncertainty is often an inherited (or 'fossilised') Type A uncertainty.

8.4.2 *Mathematical description and properties of the Gaussian distribution*

A Gaussian probability distribution is fully specified by two parameters: the mean, μ, and the variance, σ^2 (or, equivalently, the standard deviation, σ). If x is distributed as a Gaussian variable, with mean μ and variance σ^2, the probability density, $p(x)$, of x has the form (Devore 2003)

$$p(x) = \frac{1}{\sigma\sqrt{2\pi}}e^{-(x-\mu)^2/(2\sigma^2)}. \tag{8.12}$$

The factor $1/(\sigma\sqrt{2\pi})$ ensures that

$$\int_{-\infty}^{+\infty} p(x)\,\mathrm{d}x = 1. \tag{8.13}$$

It may also be shown that

$$\int_{-\infty}^{+\infty} xp(x)\,\mathrm{d}x = \mu \tag{8.14}$$

and

$$\int_{-\infty}^{+\infty} x^2 p(x)\,\mathrm{d}x - \mu^2 = \sigma^2. \tag{8.15}$$

Equations (8.14) and (8.15) verify that μ and σ^2 are in fact the mean and variance, respectively, of the Gaussian population.

We note the following features of the general shape in figure 8.5. The curve is symmetric about its peak, but declines steeply as we move away from the peak. The peak value is also the mean, in view of the symmetry of the curve about the peak. One standard deviation (1σ) away from the mean, to the right or left, is the point of inflection of the curve, that is, where the rate of change of the gradient of the curve is zero.

Between the two one-standard-deviation (1σ) points, on either side of the peak, is 68% of the total area under the curve. Between the two two-standard-deviation (2σ) points (more exactly, the 1.96σ points) is 95% of the total area of the curve. This 95% fraction plays an important role in metrology, since we often speak of a 'level of confidence' of 95% that the true value of a measurand lies between two stated limits, and these are, approximately, the $\pm 2\sigma$ points. There is

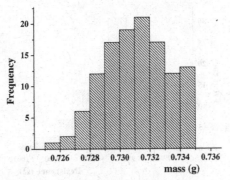

Figure 8.7. Mass (g) of steel metric M3 10-mm screws in a single batch: mean 0.731 g, standard deviation 0.002 g.

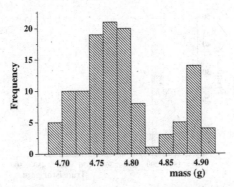

Figure 8.8. Mass (g) of steel 5/16-inch nuts in a single batch: mean 4.786 g, standard deviation 0.060 g.

no bound to the Gaussian distribution; it extends from minus infinity to plus infinity. However, beyond $\pm 3\sigma$ from the mean, the area under the curve is small ($<0.3\%$).

8.5 Experimentally observed non-Gaussian distributions

Figures 8.7–8.10 illustrate likely cases of non-Gaussian distributions. In figure 8.7, which shows the distribution of mass of steel screws packaged in one box, the distribution is truncated so that masses above a particular value appear to be missing. This could be a result of quality control following manufacture, when sizes (and therefore masses) of screws above a predetermined value were automatically discarded. In figure 8.8, the steel nuts appear to have been manufactured in two lots (perhaps using different machines or by different personnel), although they were all packaged in one box.

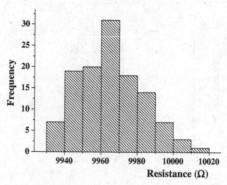

Figure 8.9. Resistance (Ω) of 0.25-W, 10-kΩ metal-film resistors in a single batch: mean 9965.47 Ω, standard deviation 17.23 Ω.

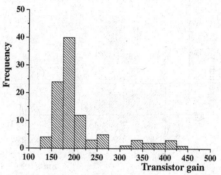

Figure 8.10. BC107 transistor gain h_{fe}: mean 209.4, standard deviation 66.9.

8.5.1 The lognormal distribution

Figures 8.9 and 8.10 show the observed distributions of samples of components used in electronics: resistances of 0.25-W metal-film resistors, of nominal value 10 kΩ, in figure 8.9, and current gains of BC107 transistors in figure 8.10. The shapes of these distributions suggest the so-called 'lognormal' shape, whose probability density distribution is illustrated in figure 8.11(a). This distribution is characterised by a steep rise towards the peak, followed by a shallow, long and exponentially decreasing tail. A variable, x, is said to have a lognormal distribution if $\log x$ has a normal or Gaussian distribution; hence the name 'lognormal', and the Gaussian distribution corresponding to figure 8.11(a) is shown in figure 8.11(b).

The Gaussian distribution was described above as arising from the additive combination of small random influences. The lognormal distribution arises from the *multiplicative* combination of small random influences. Since the logarithm of a product of terms is the sum of their logarithms, it can be shown that, if x is

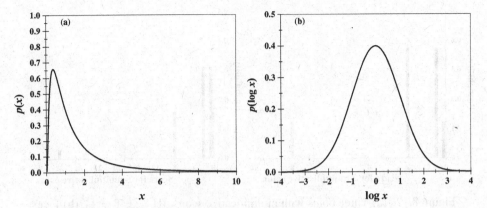

Figure 8.11. (a) A typical probability density distribution of lognormal variable x. (b) The Gaussian density distribution of $\log x$.

lognormal, being the net *product* of a number of influences, then $\log x$ is a *sum* of a set of random influences and is Gaussian.

Many natural and artificial phenomena are distributed roughly lognormally: growth of bacteria, frequency of rainfall, annual personal income, stockmarket prices, corrosion in metal structures and variations in artefact standards used in metrology.[10]

We now show how the multiplicative combination of small random influences creates the steep rise to the peak and the long thinly populated tail of the lognormal distribution. Suppose that three fair coins are tossed and that the scores (equivalent to small influences) are 2 for heads and $\frac{1}{2}$ for tails, and the total score is the *product* of the three individual scores. Then the eight outcomes will be

$$
\begin{array}{ll}
\text{HHH} & 8 \\
\text{HHT} & 2 \\
\text{HTH} & 2 \\
\text{HTT} & \frac{1}{2} \\
\text{THH} & 2 \\
\text{THT} & \frac{1}{2} \\
\text{TTH} & \frac{1}{2} \\
\text{TTT} & \frac{1}{8}
\end{array}
$$

Out of eight possible outcomes, a score of, for example, 2 is obtained three times and so has probability $\frac{3}{8} = 0.375$. The eight probabilities are plotted in figure 8.12(a), to be compared with the additive, Gaussian case of figure 8.1(c). The case of five tossed coins with the same multiplicative scores is shown in

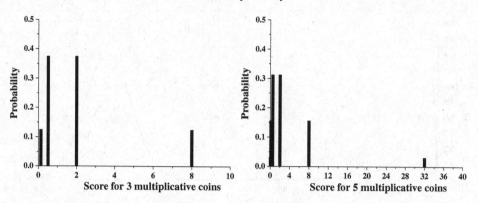

Figure 8.12. (a) Three coins with multiplicative scores (H = 2, T = $\frac{1}{2}$). (b) Five coins with multiplicative scores (H = 2, T = $\frac{1}{2}$).

figure 8.12(b), to be compared with figure 8.1(d). The steep rise to the peak and the long 'tail' are already in evidence in figures 8.12(a) and 8.12(b).

Just as the Gaussian shape does not depend on the choice of the individual scores, as long as they are additive, neither does the lognormal shape, as long as they are multiplicative. The analogue of a constant K or $-K$ as the individual scores for the Gaussian case (in figures 8.1(a)–(f) we had $K = 1$) is a factor C or $1/C$ for the lognormal case (in figures 8.12(a) and (b), $C = 2$). We note that C and $1/C$ have the same sign.

When influences combine in a multiplicative fashion, we may regard any change in a lognormal variable, resulting from an influence, as proportional to the existing magnitude of the variable. The change may be such as to increase or decrease the magnitude. The population of microorganisms such as bacteria or a fungus in a particular plant species is likely to be lognormally distributed, since, if the existing amount of microorganism is x, the rate of change is Cx.

Such a multiplicative process may take place in manufactured goods, including, for example, the resistors and transistors in figures 8.9 and 8.10, for which a roughly lognormal distribution seems to be present.[11] In the case of the transistors, in particular, we see that the process need not necessarily entail the propagation of a 'defect', since in the tail of the distribution we have transistors of unusually high gain for the type number. For many applications, high gain is desirable. However, in metrology the same process in artefact standards usually has undesirable results. Artefact standards that realise a particular unit or multiple of a unit (for example, a 500-g standard weight, a 10-V voltage standard or a platinum resistance

---

[11] The shape of a histogram is sensitive to the bin size, so we should be cautious about inferring a particular distribution from a single histogram (whose bin size may be automatically selected by the software used for creating the histogram). There are objective tests for determining how well an observed distribution fits a theoretical distribution. One such test is the 'chi-square goodness of fit test' (Bendat and Piersol 2000).

thermometer for a specified temperature range) are manufactured with meticulous care and should be identical to other artefact standards of the same nominal value. Nevertheless, their exact values differ and often have a lognormal distribution.

8.5.2 Truncated Gaussian distributions

A distribution that would otherwise be Gaussian may be truncated at some physically imposed limit. Thus, angles measured in coordinate metrology cannot be negative, and in chemical metrology the purity of an element or compound cannot exceed 100%. If the variable has values very close to a physically imposed limit, we must assume truncation at that limit. 'Very close' implies that the quantity being measured has a mean and standard deviation that together bring it to a physically imposed limit. The contrasting case arises where such a limit is many standard deviations distant from the mean; a Gaussian distribution is then possible, to a very good approximation. Such an example is provided by the histogram of masses of fruit in figure 8.6; the fact that mass cannot be negative has no effect on the shape of the histogram.[12]

8.6 The central limit theorem

If non-Gaussian distributions occur regularly, does this invalidate the application of the Gaussian distribution in the determination of uncertainties in measurement? The central limit theorem predicts that a Gaussian distribution will result (usually to a good approximation) when we calculate the sums, and therefore means, of samples whose elements are *randomly* drawn from non-Gaussian distributions.[13] Calculating the mean is the most common operation carried out on experimental data, and so the central limit theorem in effect restores and validates the Gaussian assumption.

Figures 8.1 and 8.2 show, respectively, the variation in the shape of the discrete distribution of scores using coins and dice. For a single coin or die, the distribution is the discrete equivalent of the uniform distribution discussed in section 8.3. As the number of coins or dice increases, the shape of the distribution of the sum approaches the Gaussian distribution. We may now ask the obvious question regarding the *continuous* counterpart of these discrete distributions: if we draw at random two, three, four or more elements from a continuous uniform distribution and add them together, what is the distribution of the sum?

---

[12] A Gaussian distribution that would have zero mean if untruncated, but is truncated at its peak to have only positive values, is shown in figure 8.16(a) later.

[13] In this chapter, we refer to the individual items in a sample as its 'elements'. Each element has a numerical value, so that we can calculate the sum and mean of these values. 'Randomly drawn' implies that all the values in a sample are obtained independently of one another.

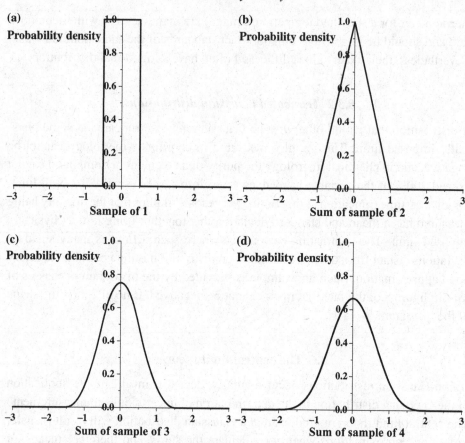

Figure 8.13. Probability density distributions of sums of samples consisting of one, two, three and four elements from a uniform distribution.

As we might predict from figures 8.1 and 8.2, the sum of two elements drawn at random from a uniform distribution is distributed as a triangular distribution. When we draw more than two elements at random from the uniform distribution, their sum approaches the Gaussian distribution, as shown in figures 8.13(a), (b), (c) and (d) for the sum of one, two, three and four randomly drawn elements, respectively, showing a progressive trend towards a Gaussian distribution.[14]

The tendency for the distributions of sums and means of samples taken from a distribution to become more nearly Gaussian as the sample size increases is a prediction of the central limit theorem.

We shall give several examples of approaches to the Gaussian distribution. Although a distribution, on its way towards the Gaussian shape, may change in

---

[14] It can be shown that, as the number of randomly drawn elements from the uniform distribution increases, the distribution of the sum of the elements is composed of a large number of high-order smoothly joined polynomial curves whose combined extent increases until it becomes a Gaussian extending from $x = -\infty$ to $x = +\infty$.

complicated ways, a simple and useful relationship holds between the mean of the distribution of the sum of a randomly drawn sample and the means of the component distributions that provide the individual elements of that sample. A similar relationship holds for the respective variances.[15] These relationships may be stated as follows.

8.6.1 Distribution of the sum of a sample

Suppose that each individual element, z_i ($i = 1, 2, \ldots, n$), of a sample of size n is *randomly* drawn from a population with its own probability density distribution, D_i. (The population may be a different one for each element.) Let μ_i be the mean and σ_i^2 the variance of D_i. We calculate the sum $S = \sum_{i=1}^{n} z_i$ of this sample of size n. The sum, S, will have its own probability density distribution, D_S. Then the mean of D_S is $\mu_1 + \mu_2 + \cdots + \mu_n$ and the variance of D_S is $\sigma_1^2 + \sigma_2^2 + \cdots + \sigma_n^2$. Here are some examples for the particular case where the D_i are all the *same* distribution (this being the case when we have a single distribution and randomly draw samples of varying size from it alone).

We start with the uniform distribution of half-width $\frac{1}{2}$ in figure 8.13(a). The above relationships yield the following results. The means of the distributions in figures 8.13(b)–(d) are all zero, since the mean of the distribution in figure 8.13(a) is zero, and this result is obvious from the symmetry in figures 8.13(b)–(d). Since the variance of the uniform distribution is $\frac{1}{12}$ (equation (8.10) with $a = \frac{1}{2}$), the variances of the distributions in figures 8.13(b)–(d) are, respectively, $2 \times \frac{1}{12} = \frac{1}{6}$, $3 \times \frac{1}{12} = \frac{1}{4}$ and $4 \times \frac{1}{12} = \frac{1}{3}$ (standard deviations respectively $\sqrt{\frac{1}{6}} \simeq 0.41$, $\frac{1}{2}$ and $\sqrt{\frac{1}{3}} \simeq 0.58$).

Next, we consider a quantity distributed as a one-sided exponential distribution. For this quantity,

$$p(x) = \begin{cases} e^{-x}, & x \geq 0, \\ 0, & x < 0. \end{cases} \tag{8.16}$$

It may be checked that $\int_{-\infty}^{\infty} p(x)\, dx = 1$, satisfying equation (8.3). The probability density $p(x)$ shown in figure 8.14(a) is a maximum at $x = 0$, but the mean, μ, of x is at $x = 1$. For an asymmetrical distribution such as this, the locations of the maximum and the mean are expected to be different. Since this distribution has a long right-hand tail, μ exceeds the value that x has (namely, zero) at the peak of the

---

[15] These relationships have appeared previously under a different guise; thus the relationship for the means is simply rule (c) in section 5.1.1, and the relationship for the variances was discussed in section 7.1.1. The relationships appear in formal proofs of the central limit theorem. Proofs of the theorem may be found in chapter 7 of Kendall and Stuart (1969).

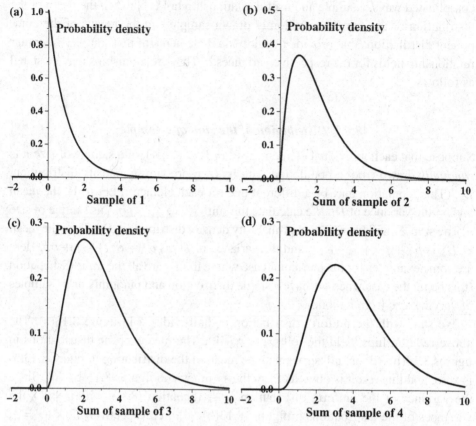

Figure 8.14. Probability density distributions of sums of samples consisting of one, two, three and four elements from a one-sided exponential distribution.

distribution. When two elements are drawn at random from this distribution, their sum is distributed as in figure 8.14(b). Perhaps contrary to intuition, the maximum of this distribution is not at $x = 0$ but at $x = 1$. Its mean is at $x = 2$, following the relationship for means stated above. With three and four elements drawn at random from the exponential distribution, the distribution of the sum moves further to the right as shown in figures 8.14(c) and 8.14(d), becoming more symmetric and approaching a Gaussian shape. The means of the distributions in figures 8.14(c) and 8.14(d) are respectively 3 and 4.

The variance of the one-sided exponential in figure 8.14(a) may be shown to be 1 (using equation (8.5)). The variances of the distributions in figures 8.14(b)–(d) are therefore respectively 2, 3 and 4 (standard deviations $\sqrt{2} \simeq 1.41$, $\sqrt{3} \simeq 1.73$ and 2), following the relationship for variances stated above.

Figure 8.15(a) shows a 'central-dip' parabolic distribution, defined by $p(x) = \frac{3}{2}x^2$ for x between -1 and $+1$ and $p(x) = 0$ elsewhere. (The factor $\frac{3}{2}$ ensures that

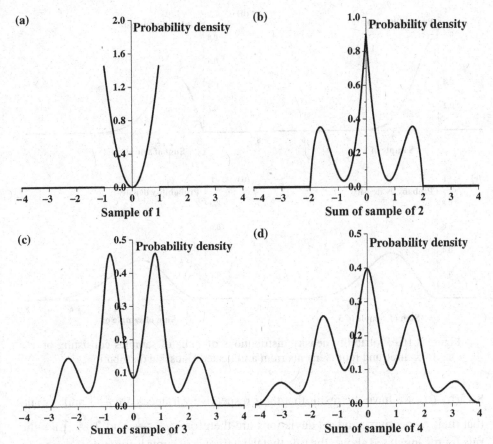

Figure 8.15. Probability density distributions of sums of samples consisting of one, two, three and four elements from a central-dip parabolic distribution.

$\int_{-1}^{+1} p(x)\,\mathrm{d}x = 1$.) With this distribution, x is more likely to take values near the extremes of its permitted range, rather than near the centre.[16] This distribution is, therefore, radically different from the Gaussian. Nevertheless, figure 8.15(b) shows how the distribution of the sum of a sample of just two elements taken from this distribution has already acquired a central peak. In figures 8.15(c) and 8.15(d), showing respectively the distributions of sums of samples consisting of three and four elements, the envelope approaches the Gaussian shape, although side-lobes are still prominent.

For this central-dip parabolic distribution in figure 8.15(a), it may be shown, using equation (8.5), that its variance is given by $\frac{3}{2}\int_{-1}^{+1} x^4\,\mathrm{d}x = \frac{3}{5}$ and the standard deviation is therefore $\sqrt{\frac{3}{5}}$. Thus, in spite of the complicated shapes of figures

---

[16] Symmetrical distributions with high densities at the edges and a low density at the centre are encountered in microwave metrology (Harris and Warner 1981).

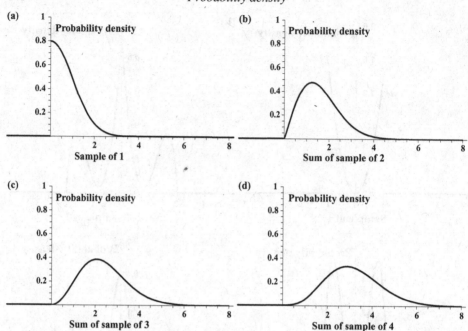

Figure 8.16. Probability density distributions of sums of samples consisting of one, two, three and four elements from a truncated Gaussian distribution.

8.15(b)–(d), we have the result that their respective variances are $\frac{6}{5}$, $\frac{9}{5}$ and $\frac{12}{5}$, and that their respective standard deviations are therefore $\sqrt{\frac{6}{5}}$, $\frac{3}{\sqrt{5}}$ and $2\sqrt{\frac{3}{5}}$. Like the rule for means stated above, the rule that the variance of sums is the sum of variances (for uncorrelated populations) is useful inasmuch as the details of the probability density distributions are not required.

Figure 8.16(a) shows a Gaussian distribution that is truncated at its peak to positive values only. If the 'full' Gaussian distribution has mean equal to 0 and standard deviation equal to 1, this truncated distribution may be defined, from equation (8.12), as

$$p_{\text{trunc}}(x) = \begin{cases} \sqrt{\dfrac{2}{\pi}}e^{-x^2/2}, & x \geq 0, \\ 0, & x < 0. \end{cases} \tag{8.17}$$

There is an extra factor of 2 in equation (8.17) compared with equation (8.12), since we require that $\int_{-\infty}^{+\infty} p_{\text{trunc}}(x)\,dx = 1$.

The mean of this truncated distribution may be shown to be $\sqrt{2/\pi} \simeq 0.798$, and its standard deviation $\sqrt{1-(2/\pi)} \simeq 0.603$. Figures 8.16(b)–(d) show respectively the distributions of sums of 2, 3 and 4 from such a truncated Gaussian distribution. Again, the distributions approach a symmetrical Gaussian distribution. The means

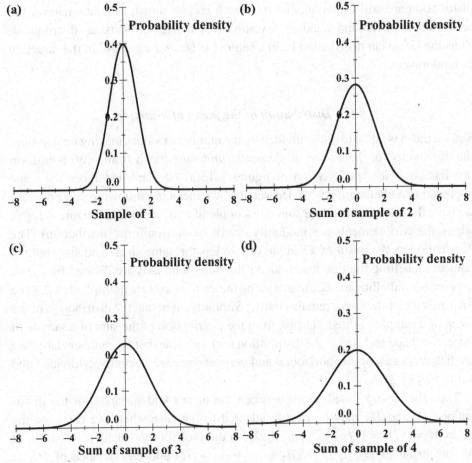

Figure 8.17. Probability density distributions of sums of samples consisting of one, two, three and four elements from a Gaussian distribution.

of the distributions of figures 8.16(b)–(d) are, respectively, $2 \times 0.798 = 1.596$, $3 \times 0.798 = 2.394$ and $4 \times 0.798 = 3.192$. The respective standard deviations are $\sqrt{2} \times 0.603 = 0.853$, $\sqrt{3} \times 0.603 = 1.044$ and $2 \times 0.603 = 1.206$.

Figures 8.17(a)–(d) show the sequence of distributions when the original distribution is Gaussian. Here the original distribution is given a mean 0 and a variance 1 (standard deviation therefore also 1). The distribution of figure 8.17(b), for the sum of a sample of two elements, has mean zero, variance 2 or standard deviation $\sqrt{2}$. The distribution of the sum of a sample of three elements (figure 8.17(c)) has mean zero, variance 3 and standard deviation $\sqrt{3}$, and the distribution of the sum of a sample of four elements (figure 8.17(d)) has mean zero, variance 4 and standard deviation 2. As figure 8.17 suggests, the distributions of sums from the original distribution are still Gaussian, and this result can be shown to hold whatever the values of the mean and standard deviation of the original Gaussian distribution. Thus the Gaussian distribution is, in a sense, 'as far as we can go' in the direction of randomness.[17]

8.6.2 Distribution of the mean of a sample

Since a mean is equal to a sum divided by the number of values making up that sum, the distribution of the *means* of elements *randomly* drawn from (say) a uniform distribution is a scaled version of figures 8.13(a)–(d), and undergoes the same approach to a Gaussian. By 'scaled version' we mean the following. Suppose that we have the distribution of the sum of a sample of two (for example figure 8.13(b), where the two elements are randomly drawn from a uniform distribution). The distribution of the mean of a sample of two has the same shape and size, but the numbers labelling the tick marks along the horizontal axis are divided by 2, and the numbers labelling the tick marks along the vertical axis are multiplied by 2. (The total area under the curve remains unity.) Similarly, to obtain the distribution of the mean of a sample of three, starting from the distribution of the sum of a sample of three, the shape and size of the distribution stay the same, but the numbers labelling the tick marks along the horizontal and vertical axes are respectively divided and multiplied by 3.

The relationship stated above between the means and variances of the distributions D_i and D_S may be readily adapted to the case where we calculate the mean $M = S/n = (1/n)\sum_{i=1}^{n} z_i$. If D_M is the probability density distribution of M, the mean of D_M is $(1/n)(\mu_1 + \mu_2 + \cdots + \mu_n)$ and the variance of D_M is $(1/n^2)(\sigma_1^2 + \sigma_2^2 + \cdots + \sigma_n^2)$. If the sample elements are randomly drawn from the *same* distribution, so that $\sigma_1^2 = \sigma_2^2 = \cdots = \sigma_n^2 = \sigma^2$, this rule for variances implies that the variance of D_M is $(1/n^2)n\sigma^2 = \sigma^2/n$. This is a restatement of (for example) equation (5.56). We recall, from previous discussions, that the values of a sample of size n, drawn from a population with variance σ^2, must be uncorrelated if the variance of the mean of that sample is σ^2/n. In the present context, we see that it is the *randomness* of draws from a population that provides the necessary absence of correlation.

We may now express the central limit theorem as follows. Suppose that we make n independent measurements of a non-Gaussian random variable, x, and we calculate their mean, $\bar{x}$. Let x have a population mean μ and a population variance

---

[17] We should note, however, that a sequence of readings may present significant autocorrelation, yet may also have a Gaussian distribution. A Gaussian distribution of serial readings therefore does not necessarily imply 'white noise' (this term was introduced in section 7.2.2).

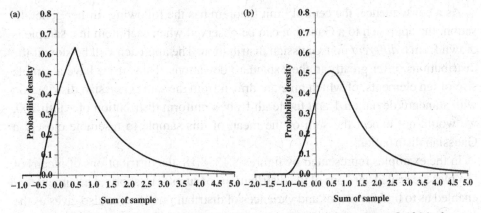

Figure 8.18. (a) The sum of a sample of two, one element from figure 8.13(a), the other from figure 8.14(a). (b) The sum of a sample of three, two elements from figure 8.13(a), the third from figure 8.14(a).

σ^2. Then the distribution of $\bar{x}$ approaches a Gaussian distribution as n increases,[18] and this Gaussian distribution has mean μ and variance σ^2/n. As indicated in chapter 5, we can estimate μ unbiasedly as $\bar{x}$ (equation (5.2)), and we can estimate σ^2 unbiasedly using s^2 as in equation (5.8).

The relationship stated above between D_S and D_i and between D_M and D_i remain valid when the D_i ($i = 1, 2, \ldots, n$) are different distributions. Suppose that we take a sample consisting of two elements, one drawn at random from the uniform distribution of figure 8.13(a) and the other from the one-sided exponential distribution of figure 8.14(a). We calculate the sum of these two elements. Its distribution is shown in figure 8.18(a). The mean and variance of the distribution in figure 8.13(a) are respectively 0 and $\frac{1}{12}$, and the mean and variance of the distribution in figure 8.14(a) are respectively 1 and 1. Hence the mean of the distribution in figure 8.18(a) is $0 + 1 = 1$, and the variance of this distribution is $\frac{1}{12} + 1 = \frac{13}{12}$ (standard deviation $\sqrt{\frac{13}{12}} \simeq 1.04$). Figure 8.18(b) shows the distribution of the sum of three elements, two drawn from the uniform distribution and one from the one-sided exponential distribution. As expected, this distribution is smoother and more symmetrical than that in figure 8.18(a). The mean of this distribution is $0 + 0 + 1 = 1$, and the variance of this distribution is $\frac{1}{12} + \frac{1}{12} + 1 = \frac{7}{6}$ (standard deviation $\sqrt{\frac{7}{6}} \simeq 1.08$).

---

[18] There are distributions, such as the Cauchy distribution (Bevington and Robinson 2002), where the approach to a Gaussian does not take place no matter how large the sample. Such distributions are not commonly encountered in metrology.

As a consequence, the central limit theorem has the following further generalisation: the approach to a Gaussian can be observed when each item in a sample is drawn from a *different* non-Gaussian distribution. The approach will be slow if the distributions differ greatly in their standard deviations. Thus, if we have a sample size of ten elements, of which nine are drawn from the same Gaussian distribution with standard deviation 1 and the tenth from a uniform distribution of width 100, we would not expect the sum or the mean of this sample to resemble closely a Gaussian distribution.

In the examples represented by figures 8.13–8.18, the distributions of means of samples are scaled versions of the distributions of sums. The shortcut argument that enabled us to find the means and variances of distributions of sums also gives us the means and variances of the distributions of means. We may illustrate this starting from the uniform distribution of figure 8.13(a). Since the triangular distribution in figure 8.13(b) of the sum of a sample of two elements drawn from a uniform distribution has a variance of $\frac{1}{6}$, the distribution of the mean of a sample of two elements from a uniform distribution has a variance $\frac{1}{4} \times \frac{1}{6} = \frac{1}{24}$ or standard deviation $\sqrt{\frac{1}{24}}$ or $\frac{1}{2\sqrt{6}}$. The standard deviation of the distribution of the mean of a sample of size two from the uniform distribution of figure 8.13(a) is, therefore, less by $\sqrt{2}$ than the standard deviation $\sqrt{\frac{1}{12}}$ of the distribution consisting of samples in which each sample consists of a single element from the same uniform distribution. This recalls the fact that, if x_1 and x_2 are uncorrelated values from the same population with variance σ_x^2, then the sum $x_1 + x_2$ has variance $2\sigma_x^2$ and the mean $\frac{1}{2}(x_1 + x_2)$ has variance $\frac{1}{4} \times 2\sigma_x^2 = \frac{1}{2}\sigma_x^2$. So, if x_1 and x_2 each has a standard deviation σ_x, their mean $\frac{1}{2}(x_1 + x_2)$ has a standard deviation $\sigma_x/\sqrt{2}$. This is a restatement of (for example) equation (5.56) in chapter 5 with $n = 2$.

Exercise D

(1) The probability density for a particular distribution is given by $p(x) = Ax^4$ for $-1 < x < +1$. For other values of x, $p(x) = 0$.
   (a) For this probability density, show that the value of the constant $A = \frac{5}{2}$.
   (b) Calculate the mean and standard deviation of the distribution.
   (c) What is the probability that x lies between -0.5 and $+0.5$?
   (d) Calculate the mean and standard deviation of the distribution of the mean of samples of six values drawn from this distribution.

(2) The probability density for a particular distribution is given by $p(x) = 1$ for $0 < x < +1$. For other values of x, $p(x) = 0$.
   (a) For this probability density, calculate the mean and standard deviation of the distribution.

(b) Calculate the mean and standard deviation of the distribution of the mean of samples of two values drawn from this distribution.

   Use the uniform random-number generator[19] on a spreadsheet to generate 2000 numbers in the interval 0 to 1. Taking these numbers in pairs, calculate the mean of each pair and create a column consisting of 1000 means.

(c) Calculate the mean and standard deviation of the 1000 means – compare this with your answer for part (b).

8.7 Review

The examples shown in figures 8.13–8.17 are instances of the central limit theorem in operation. Although in these examples we considered sums (and means) taken from the *same* distribution, the approach to a Gaussian distribution also takes place if each of the elements in the sample is drawn at random from a *different* distribution, as in figure 8.18. This is the essence of the central limit theorem. The approach to a Gaussian will be gradual or even very slow if one or several of the component non-Gaussian distributions have a much larger standard deviation than the others. However, in most cases the distribution of a measurand y, which is the sum of inputs $y = x_1 + x_2 + \cdots + x_n$, may be considered Gaussian (or at least approximately so) when some or all of the inputs x_i are non-Gaussian. This finding also holds for measurands that are more complicated functions of the inputs x_i, and explains the great metrological usefulness of the theorem.

   In the next chapter we will consider in more detail how the properties of a sample (such as the mean, variance and standard deviation) drawn from a Gaussian distribution are affected as the size of the sample changes.

---

[19] The function RAND( ) in Excel will generate numbers in the interval 0 to 1 with uniform probability.

9

Sampling a Gaussian distribution

If it is reasonable to assume that a population consists of values that have a Gaussian distribution, then what will be the distribution of a property (a 'statistic') of a *sample* drawn from this Gaussian 'parent'? The property might be the mean, variance or standard deviation of the sample. Each of these properties has a *sampling distribution*, which can be described as follows.

We imagine a very large or infinite population that has a Gaussian distribution with mean μ and standard deviation σ. A sample consisting of n values is *randomly* drawn from this population. A property of the sample is calculated, in order to estimate the corresponding population parameter. We then draw another sample, also of size n, and calculate the same property for this second sample. The process is repeated many times. Next the distribution of that property is examined; the distribution becomes manifest as a result of taking a *large* number of repeated samples (all of size n). The distribution is the sampling distribution of the property in question. It is understood that, in any particular experimental situation, we do not actually need to draw a large number of samples; this process is a conceptual one that enables us to infer, from *one* actual sample, the variability (depicted by the shape of the sampling distribution) of our estimate of the population parameter. In section 9.1 we review the material already discussed in section 8.6.2.

9.1 Sampling distribution of the mean of a sample of size n, from a Gaussian population

Assume a Gaussian population with mean μ and variance σ^2. Let x_i $(i = 1, 2, \ldots, n)$ be a value in a sample of size n randomly drawn from the population. We discovered in chapter 5 that, in terms of expectations, μ and σ^2 may be expressed as $E(x_i) = \mu$ and $E(x_i^2) - \mu^2 = \sigma^2$.

The mean, $\bar{x}$, of the sample is given by

$$\bar{x} = \frac{x_1 + x_2 + \cdots + x_n}{n}.$$

The sampling distribution of $\bar{x}$ itself has a mean given by[1]

$$E(\bar{x}) = \frac{1}{n}[E(x_1) + E(x_2) + \cdots + E(x_n)]$$

$$= \frac{1}{n}[\mu + \mu + \cdots + (n \text{ times})] = \frac{1}{n}n\mu = \mu. \tag{9.1}$$

We conclude that, whatever the shape of the distribution of the means $\bar{x}$ of samples of size n, the mean of this distribution must be μ, like the mean of the parent distribution.

The variance, $\sigma_{\bar{x}}^2$, of the distribution of the means of samples of size n is given by

$$\sigma_{\bar{x}}^2 = \frac{\sigma_x^2}{n}, \tag{9.2}$$

where σ_x^2 is the variance of each value, x_i, in the sample; thus $\sigma_x^2 = \sigma^2$, the variance of the parent population to which each such value belongs. Equation (9.2) is valid when the x_i are values randomly drawn from the parent population. The standard deviation, $\sigma_{\bar{x}}$, of the distribution of the means of the samples of size n is, therefore, from equation (9.2),

$$\sigma_{\bar{x}} = \frac{\sigma_x}{\sqrt{n}}. \tag{9.3}$$

Figure 9.1 shows the shapes of the sampling distribution of $\bar{x}$ for $n = 1,\ 4,\ 10$ and 20 when the parent population is Gaussian with mean $\mu = 0.3$ and standard deviation $\sigma = 1$. The shapes are all Gaussian; this preservation of the Gaussian shape, when samples are drawn at random from a Gaussian parent and the sums or means of these samples are calculated, is shown in figures 8.17(a)–(d). The larger the sample size, the more reliable the estimate of the population mean, as is shown by the narrower Gaussian curves for samples of larger n.

9.2 Sampling distribution of the variance of a sample of size n, from a Gaussian population

A sample $(x_1,\ x_2, \ldots, x_n)$ of size n and mean $\bar{x}$ provides an unbiased estimate, s^2, of the population variance given by

$$s^2 = \frac{1}{n-1}[(x_1 - \bar{x})^2 + (x_2 - \bar{x})^2 + \cdots + (x_n - \bar{x})^2]. \tag{9.4}$$

---

[1] See rules (b) and (c) in section 5.1.1, or section 8.6.2.

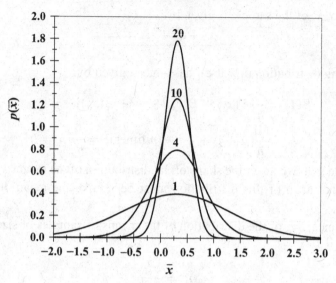

Figure 9.1. The probability density for the mean, $\bar{x}$, of samples of size $n = 1, 4$, 10 and 20 from a Gaussian parent of $\mu = 0.3$ and $\sigma = 1$.

When considering the sampling distribution of the variance we exclude the case $n = 1$, since the variance of a sample of size equal to 1 has no meaning. In addition, the variance of a sample must be zero or positive; therefore, the distribution of s^2 cannot be Gaussian, since a Gaussian variable extends from minus to plus infinity, no matter what its mean or standard deviation.

In equation (9.4), the variance, s^2, is calculated for $n - 1$ degrees of freedom. The more general expression for s^2 is[2]

$$s^2 = \frac{\sum_{i=1}^{n} \epsilon_i^2}{\nu},$$ (9.5)

where ν is the number of degrees of freedom. When we calculate the mean, $\bar{x}$, of a sample of size n, the number of degrees of freedom is $\nu = n - 1$, and this divisor, $n - 1$, appears in equation (9.4). In some situations, we may wish to obtain estimates of an intercept and a slope by fitting a straight line to n values. In cases such as these where we extract two estimates from the sample, the unbiased estimate of the population variance is calculated as the residual sum of squares divided by $n - 2$. Here the n residuals, ϵ_i, are constrained by two equations: $\sum_{i=1}^{n} \epsilon_i = 0$ and a second equation that includes the explanatory variable.[3]

The sampling distribution of s^2 depends more directly on the number of degrees of freedom than on the sample size. The distributions of s^2 for degrees of freedom

---

[2] Repeating equation (5.23).
[3] See section 5.2.3.

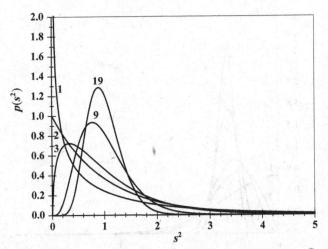

Figure 9.2. The probability density for the unbiased estimate, s^2, of variance of a Gaussian population with $\sigma^2 = 1$, for 1, 2, 3, 9 and 19 degrees of freedom.

equal to 1, 2, 3, 9 and 19 drawn from a Gaussian parent of arbitrary mean and standard deviation equal to 1, and variance therefore also equal to 1, are shown in figure 9.2.[4] For a general Gaussian parent of variance σ^2, the distributions of s^2 would be identical to those in figure 9.2, with the numerical values of probability density along the vertical axis divided by σ^2, and the numerical values along the horizontal axis multiplied by σ^2.

The distributions for ν equal to 1 and 2 in figure 9.2 peak at a variance of zero (where the probability density is infinite for $\nu = 1$), while the distributions for higher numbers of degrees of freedom peak at non-zero values of variance. All the distributions of the sample variance in figure 9.2 have a mean equal to 1, and this is true no matter what the sample size, as long as the variance of the Gaussian parent is equal to 1. The reason is that $E(s^2) = \sigma^2 = 1$, the stated variance of the population: s^2 is the unbiased estimate of the variance. More generally, the unbiased estimate of the population variance, σ^2, is given by equation (9.5), implying that

$$E(s^2) = \sigma^2, \tag{9.6}$$

which is equal to 1 in the case of figure 9.2.[5]

In figure 9.2, the greater the number of degrees of freedom, the narrower the distribution and the closer the approximation to a Gaussian shape. In general, the

---

[4] These values of the number of degrees of freedom are chosen because, when only the mean is estimated ($\nu = n - 1$ or $n = \nu + 1$) the actual sample sizes are 2, 3, 4 and the round numbers 10 and 20. The equations describing the probability densities in figures 9.2 and 9.3 are derived in Wilks (1962).

[5] It is worth noting that the unbiased property $E(s^2) = \sigma^2$, where s is calculated using equation (9.4) or equation (9.5), does not require the parent distribution to be Gaussian.

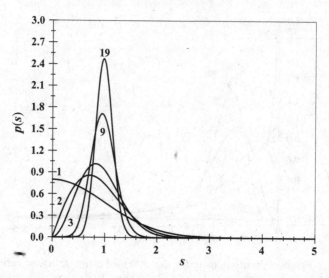

Figure 9.3. The probability density for sample standard deviation for $\nu = 1, 2, 3, 9$ and 19 degrees of freedom from a Gaussian population with $\sigma = 1$.

larger the sample size (for a given number of parameters to be estimated), the more reliable is the estimate of the population variance. The sampling distribution of the variance can *itself* be characterised by a variance, which we call $u^2(s^2)$. It can be shown that (Frenkel 2003)

$$u^2(s^2) = \frac{2\sigma^4}{\nu}. \tag{9.7}$$

Thus the higher ν, the smaller $u^2(s^2)$; hence the narrower curves in figure 9.2 for higher degrees of freedom. We note the dependence on σ^4, which is dimensionally correct, since the left-hand side of equation (9.7) is essentially the variance of a variance, namely a fourth-order term. It follows that both the left- and the right-hand side of equation (9.7) are of fourth order.[6]

The variance, s^2, plotted along the horizontal axis in figure 9.2 is related through a change of scale to a variable known as the 'chi-squared' variable for ν degrees of freedom and denoted by χ_ν^2. The definition of χ_ν^2 is

$$\chi_\nu^2 = \frac{\sum_{i=1}^n \epsilon_i^2}{\sigma^2} = \frac{\nu s^2}{\sigma^2}, \tag{9.8}$$

so that the mean of χ_ν^2 is

$$E\left(\chi_\nu^2\right) = \frac{\nu E\left(s^2\right)}{\sigma^2} = \frac{\nu \sigma^2}{\sigma^2} = \nu, \tag{9.9}$$

[6] Note that ν is dimensionless.

and the variance $u^2(\chi_v^2)$ of χ_v^2 is, using (for example) equations (7.18) and (9.7),

$$u^2(\chi_v^2) = \frac{v^2 u^2(s^2)}{\sigma^4} = \frac{v^2 2\sigma^4}{\sigma^4 v} = 2v. \tag{9.10}$$

The standard uncertainty $u(\chi_v^2)$ of χ_v^2 is therefore $\sqrt{2v}$.

The probability density graph of χ_v^2, for a given value of v, is identical to the graph in figure 9.2 for that particular value of v, with the horizontal axis marked in units $0, v, 2v, 3v, \ldots$ instead of $0, 1, 2, 3, \ldots$ The chi-squared variable is used when experimental and theoretical probability density distributions are being compared; a significantly high value of χ_v^2 (meaning a value well to the right of the peaks in figure 9.2) implies that an experimentally derived distribution is in conflict with theory.[7]

9.3 Sampling distribution of the standard deviation of a sample of size n, from a Gaussian population

The standard deviation, s, is defined as the square root of s^2 in equation (9.4):

$$s = \sqrt{\frac{1}{v}[(x_1 - \bar{x})^2 + (x_2 - \bar{x})^2 + \cdots + (x_n - \bar{x})^2]}. \tag{9.11}$$

The sampling distributions of s for $v = 1, 2, 3, 9$ and 19, drawn from a Gaussian parent of arbitrary mean and standard deviation equal to 1, are illustrated in figure 9.3. They are similar to the distributions of s^2 in figure 9.2, although for $v = 1$ the probability density is now finite, and there is a further difference: although s^2 is an unbiased estimate of the population variance σ^2, so that $E(s^2) = \sigma^2$, it does *not* follow that $E(s) = \sigma$. Thus, although in figure 9.2 s^2 for each number of degrees of freedom has a mean value equal to 1, in figure 9.3 the standard deviation, s, for each number of degrees of freedom does not have a mean equal to 1. However, the difference from 1 is small, especially for a large number of degrees of freedom; thus the means $E(s)$ of the curves for $v = 1, 2, 3, 9$ and 19 are, respectively, 0.798, 0.886, 0.921, 0.973 and 0.987, so that, as the number of degrees of freedom increases, $E(s)$ tends to σ (equal to 1 in this case) asymptotically from below.

9.3.1 The 'uncertainty of an uncertainty' and its relationship to degrees of freedom

The variance, $u^2(s)$, of the curves in figure 9.3 is given approximately by

$$u^2(s) = \frac{\sigma^2}{2v}. \tag{9.12}$$

---

[7] For a discussion of the chi-squared distribution, see Blaisdell (1998).

It follows that the standard deviation, $u(s)$, of s is given by

$$u(s) = \frac{\sigma}{\sqrt{2v}}.$$ (9.13)

If, for example, $\sigma = 1$ and $v = 9$, equation (9.13) gives approximately $u(s) \simeq 0.24$, and the near-Gaussian curve for $v = 9$ in figure (9.3) shows that $u(s) \simeq 0.24$ is a plausible value for its standard deviation.

Equation (9.7), for the variance of the variance, is exact (for Gaussian parent populations), but the above equations (9.12) and (9.13) for the variance and standard deviation of the standard deviation are only approximate. The relationship between $u^2(s^2)$ and $u^2(s)$ can be approximately derived using equation (7.14). Since $\partial s^2 / \partial s = 2s$, we have from equation (7.14) that

$$u^2(s^2) = \left(\frac{\partial s^2}{\partial s}\right)^2 u^2(s) = 4s^2 u^2(s),$$ (9.14)

and so, on substituting into the left-hand side of equation (9.14) from equation (9.7),

$$\frac{2\sigma^4}{v} = 4s^2 u^2(s),$$ (9.15)

so that

$$u^2(s) = \frac{1}{2} \frac{\sigma^4}{vs^2}.$$ (9.16)

If we approximate $s^2 \simeq \sigma^2$, equation (9.16) gives

$$u^2(s) = \frac{s^2}{2v},$$ (9.17)

agreeing with equation (9.12).

Equation (9.17) may be expressed in terms of v:

$$v = \frac{1}{2} \frac{s^2}{u^2(s)}.$$ (9.18)

Equation (9.18) has the following practical application. It is sometimes necessary to assign degrees of freedom to an uncertainty obtained from a Type B evaluation, under the circumstance in which no repeated values are available.[8] We rewrite equation (9.18) as

$$v = \frac{1}{2}\left(\frac{u(s)}{s}\right)^{-2} = \frac{1}{2}\left(\frac{u(u)}{u}\right)^{-2},$$ (9.19)

---

[8] If there existed a record of n repeated values, then n could be related to the number of degrees of freedom, v, by an equation such as $v = n - 1$ for the situation where one parameter, namely the mean, is estimated.

replacing s by the equivalent, u, which is more suited to the metrological context of evaluation of uncertainty. We can now recognise that $u(u)/u$ is the *proportional uncertainty* in our Type B-evaluated uncertainty, u. This proportional uncertainty can often be estimated (or, sometimes, frankly only guessed at). Then the appropriate degrees of freedom are given by equation (9.19). If our Type B-evaluated uncertainty has itself a proportional uncertainty of about 20%, equation (9.19) implies that about 12 degrees of freedom are associated with it. It is important to note the kind of information conveyed by the number of degrees of freedom in a measurement: it does *not* denote the uncertainty of the result, but the 'uncertainty of the uncertainty' of the result. This can clearly be seen to be the case with Type A uncertainties; thus a straight line fit to only four points, giving $\nu = 2$, results in a proportional uncertainty of roughly 50% in all the uncertainties associated with this fit.

Exercise

(1) Information accompanying a solution of copper in nitric acid indicates that the amount of copper is 9.99 mg/g with a standard uncertainty of 0.02 mg/g. Past experience indicates that the uncertainty in the standard uncertainty is 10%. Use this information to determine the number of degrees of freedom associated with the standard uncertainty in the density.

(2) The number of degrees of freedom associated with the standard uncertainty in the heat capacity of a particular liquid is eight. Use this information to calculate the fractional uncertainty in the standard uncertainty.

9.4 Review

Through the process of taking many samples each consisting of n values from a population, we are able to determine the shapes of the probability distributions of important quantities such as the sample mean, variance and standard deviation. In the next chapter we apply knowledge of the distribution of sample means and variances to establish an interval that contains the true value (otherwise known as the population mean) of a quantity with a known probability. This leads quite naturally to a quantitative expression for the expanded uncertainty of a measurand.

10

The t-distribution and the Welch–Satterthwaite formula

The uncertainty that accompanies the best estimate of a measurand is usually based on fewer than 20 degrees of freedom, and sometimes fewer than 10. The reason is as follows.

For Type A evaluations of uncertainty, the number of degrees of freedom, ν, is related to the sample size, n. Thus, when calculating the mean of a sample, $\nu = n - 1$. Where measurements are made 'manually' (not under computer control), n and therefore ν are likely to be small. Where measurements are computer-controlled and the environment is sufficiently stable, it is easy to amass samples consisting of hundreds or even thousands of values from the same population. We might therefore think that the number of degrees of freedom associated with the uncertainty in the measurand is also very high. However, this is unlikely to be so, since there will probably exist systematic errors that can be corrected for but that will nevertheless leave a Type B uncertainty. Such an uncertainty is generally associated with fewer degrees of freedom. Admittedly, the estimation of a systematic error may also be based on a large number of repeated measurements. The calibration of the $3\frac{1}{2}$-digit DMM by means of simultaneous measurements with an $8\frac{1}{2}$-digit DMM in section 6.1.2 is a case in point. A large number of such measurements could in principle allow us to determine an uncertainty in the systematic error of the $3\frac{1}{2}$-digit DMM that is associated with a large number of degrees of freedom. However, the readings of the $8\frac{1}{2}$-digit DMM themselves have an uncertainty obtained from its calibration report that is likely to be based on fewer degrees of freedom.

Somewhere along every traceability chain there is likely to be a systematic error that leaves a Type B uncertainty that can only roughly be estimated.[1] This

---

[1] In the example just given, such a traceability chain extends from the $3\frac{1}{2}$-digit DMM to the $8\frac{1}{2}$-digit DMM, and then to the high-level voltage standards based on the Josephson effect in superconductors (see section 4.1.3) used to calibrate the $8\frac{1}{2}$-digit DMM. Type B uncertainties related to Josephson-effect voltage measurements include uncertainties in corrections for thermal voltages (see section 6.2).

uncertainty is, therefore, based on only a few degrees of freedom, as implied by Equation (9.19). As will be seen in the discussion of the Welch–Satterthwaite formula in section 10.3, the combining of uncertainties based on a large number of degrees of freedom with those based on a small number of degrees of freedom is likely to create a combined uncertainty with a small number of degrees of freedom. This is not surprising; it is the rough metrological analogue of the chain that is no stronger than its weakest link.

The measurand, therefore, has an uncertainty that is generally associated with a *small* number of degrees of freedom. That is why we need the *t*-distribution. We shall illustrate how this comes about by calculating a *coverage interval* for the measurand.

The best estimate of the true value of a measurand is derived from a sample drawn from a population. The coverage interval for the measurand is that interval within which the true value of the measurand is located with high probability, usually 95% or (less commonly) 99%. Very often this interval is symmetrical about the best estimate. At the end of the experiment, we would like to know the coverage interval for the measurand, since it answers the following question: 'how well have we located the true value of the measurand?'

We note here that there is a trade-off between the confidence associated with a coverage interval and what we might call 'interesting information'. Thus we could state a coverage interval that gives us a probability of 100% that the true value lies within the interval, but this interval would be of no interest! The reason is that such a coverage interval would extend over the entire theoretically permitted range of the measurand. But we already know that the measurand has this permitted range, so we have learned nothing new. By way of example, without taking any measurements we could declare, with 100% confidence, that the temperature of distilled liquid water in a beaker, at normal atmospheric pressure, is between $0\,°C$ and $100\,°C$.

10.1 The coverage interval for a Gaussian distribution

Suppose that a population has a Gaussian distribution with mean μ and standard deviation σ. We draw a sample of size n from the population and calculate its mean, $\bar{x}$. The expectation value[2] of $\bar{x}$ is μ, thus $E(\bar{x}) = \mu$. We have, therefore, an unbiased estimate of the quantity of prime interest, namely the population mean, which we take to be equal to the true value of the measurand. We also need an estimate of *how well* we know μ. Such an estimate is provided by the coverage interval. With every coverage interval there is an associated probability. Though any probability could be chosen, most metrologists adopt an interval that contains the true value

---

[2] See equation (5.3).

with a probability of 0.95. An equivalent way to express this is to refer to the 95% coverage interval, by which it is understood that, if many intervals were calculated using samples drawn from a population, those intervals would contain the true value of the measurand in (on average) 95 out of 100 occasions.

From a sample of size n, we are able to calculate an unbiased estimate, s^2, of the variance of the population, σ^2, and, using s^2, we can obtain an approximate estimate of the standard deviation, σ, of the Gaussian population.[3]

We assume that the values in the sample are mutually uncorrelated, so that the standard deviation of $\bar{x}$ is given by

$$s_{\bar{x}} = s/\sqrt{n}. \tag{10.1}$$

The sample mean, $\bar{x}$, is itself a Gaussian variable.[4] With $\bar{x}$ as an unbiased estimate of μ, and $\bar{x}$ having a variability described by its standard deviation $s/\sqrt{n}$, we may write, notionally,

$$\bar{x} = \mu \pm \frac{s}{\sqrt{n}}. \tag{10.2}$$

To answer the question 'how well do we know μ?', we interpret equation (10.2) as follows. We regard μ as having a value that is the unknown 'true' value of the measurand. However, we do not 'see' this true value as a perfectly sharp image; it is blurred or indistinct by an amount estimated as $\pm s/\sqrt{n}$ that we regard as the uncertainty in the value of μ.[5]

We assume for the present that the term $s/\sqrt{n}$ in equation (10.2) is a constant quantity. (For small sample sizes, we shall soon discover that this assumption gives unsatisfactory results.) With $s/\sqrt{n}$ a constant quantity and $\bar{x}$ a Gaussian variable, figure 10.1 shows the Gaussian distribution of $\bar{x}$, centred on μ and having standard deviation $s/\sqrt{n}$. The 95% coverage interval for $\bar{x}$ is equal to $\bar{x} \pm$ some multiple of $s/\sqrt{n}$, this multiple being chosen so that the two 'tail regions' in figure 10.1 each have an area that is 2.5% of the total area under the probability density curve. For a Gaussian distribution this multiple is approximately 1.96. For a 95% coverage

---

[3] We may, if we wish, calculate an *exact* unbiased estimate of σ. If we have three degrees of freedom, as when calculating the mean of a sample of $n = 4$ values, then $E(s) \simeq 0.921\sigma$ (see section 9.3). It follows that, for three degrees of freedom, the unbiased estimate of σ is not exactly s but rather $s/0.921 \simeq 1.086s$, because $E(1.086s) = 1.086E(s) \simeq 1.086 \times 0.921\sigma \simeq \sigma$. This refinement is not necessary for the argument being developed here.

[4] See the discussion in section 9.1.

[5] The treatment in this book is consistent with the conventional, so-called 'frequentist', statistical approach. In this approach, the sampled quantities, for example $\bar{x}$, are the variables, and the population parameters, for example μ, are fixed. A separate approach to statistical estimation is called 'Bayesian inference' and here the population parameters are regarded as variables with probability density distributions determined by a single sample. This is a branch of statistics in its own right, named after Thomas Bayes (1702–1761), who wrote the seminal papers on what is now known as conditional probability. A general overview of this field is given in Malakoff (1999). The GUM can be interpreted as having a partly Bayesian foundation (Kacker and Jones 2003).

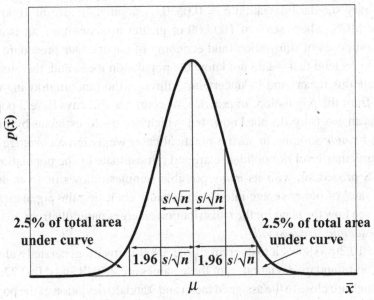

Figure 10.1. Coverage interval for the population mean, μ.

interval for μ, we therefore have

$$\bar{x} = \mu \pm 1.96 \frac{s}{\sqrt{n}}. \tag{10.3}$$

10.1.1 Using Monte Carlo simulations to study coverage intervals

Equation (10.3) will now be used to calculate the 95% coverage interval when the sample size, n, is small: specifically, $n = 4$. We perform what is known as a Monte Carlo simulation, or MCS. This technique is a kind of trial-and-error statistics, made feasible by readily available software that rapidly generates many random numbers with a specified distribution.[6] These random numbers enable us to 'simulate' a measurement process, by imparting plausible amounts of variability to the inputs to the measurand. The resulting variability in the measurand can then be observed. MCS, which can also be called 'experimental statistics', bears a relation to theoretical statistics similar to that which experimental physics does to theoretical physics.[7]

List 10.1 at the end of this chapter contains 1000 random numbers. The numbers have been generated from a Gaussian distribution with arbitrary mean $\mu = 2.5810$

---

[6] There are many commercially available software packages that generate random numbers with a specified distribution: for example, Excel, Origin and IMSL (International Mathematical Software Library).

[7] The name Monte Carlo refers to the randomness of the draw of values from the software-generated distribution, and reminds us of the mixed parentage of statistics: mathematics and gambling!

and arbitrary standard deviation $\sigma = 0.0630$. A population size of 1000 is quite small for MCS, where sizes of 100 000 or greater are common, but is adequate here for purposes of illustration (and economy of paper). Our procedure will be, in brief, to pretend that we do not know the population mean and, therefore, to try to estimate this mean (and its uncertainty) through the random drawing of small samples from the population. In practice, of course, we always have a population whose mean we truly do not know, but which we try to estimate by randomly drawing a *single* sample. In such a practical case, we assign a coverage interval with a particular level of confidence around our estimate of the population mean. The MCS procedure, with its *many* possible samples, allows us to evaluate the 'success rate' of our coverage interval in actually enclosing the population mean. We shall see how the need for the t-distribution emerges naturally from this process *when the sample size is small*.

Figure 10.2(a) shows a histogram of the 1000 software-generated values. The mean, $\bar{x}$, and standard deviation, s, of these values are 2.5818 and 0.062 77, respectively, which are close to the assigned mean and standard deviation of the population of 1000.[8] Figure 10.2(b) shows the histogram of the 250 sample means that result from drawing samples of size $n = 4$, from the population of 1000. The mean of the 250 means is 2.5818, the same to five decimal digits as the mean of the histogram of the 1000 original values. The standard deviation of the 250 means is 0.031 94, close to half the standard deviation of the 1000 original values. The narrower histogram in figure 10.2(b), compared with that in figure 10.2(a), illustrates the reduction in uncertainty by $\sqrt{n}$ (equal to 2 in this case) when a mean of n uncorrelated values is calculated. Such a reduction is the reason why we generally consider averages to be more reliable than single readings. Figures 10.2(c) and 10.2(d) are the theoretical Gaussian counterparts to figures 10.2(a) and 10.2(b), respectively.

For each of the 250 samples of size $n = 4$, drawn from the original Gaussian distribution of 1000, the standard deviation, s, can be calculated. A histogram of these 250 values of standard deviation is shown in figure 10.3(a). The mean of the 250 standard deviations is 0.057 95. For three degrees of freedom as in this case, we have[9] $E(s) \simeq 0.921\sigma$ and, since $\sigma = 0.0630$, $E(s) \simeq 0.921 \times 0.0630 \simeq 0.058\,02$, giving close agreement with the Monte Carlo-derived value of 0.057 95. Figure 10.3(b) shows a histogram of the corresponding 250 values of standard deviation of the means of the samples. The mean of these 250 standard deviations is 0.028 97, close to half the mean value of the values in figure 10.3(a). Figures 10.3(c) and 10.3(d) show the theoretical counterparts to figures 10.3(a) and 10.3(b), respectively. It can be seen that both the experimental and the theoretical

---

[8] Standard deviations are not normally stated to more than two (sometimes three) significant figures. However, for purposes of comparison of standard deviations, more figures are stated here.

[9] See section 9.3.

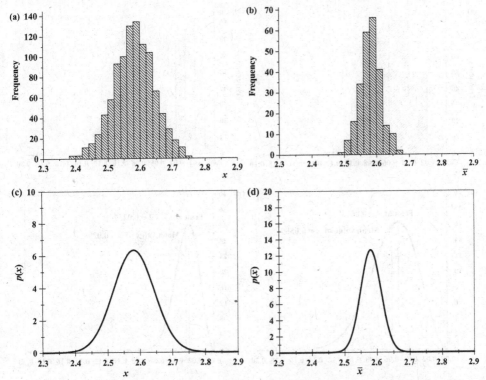

Figure 10.2. (a) A histogram of a software-generated Gaussian population of 1000 with assigned mean 2.5810 and assigned standard deviation 0.0630. The mean of the histogram is 2.5818; the standard deviation is 0.062 77. (b) A histogram of means of 250 samples of size 4 from the population shown in (a). The mean of the histogram is 2.5818; the standard deviation is 0.031 94. The mean, $\bar{x}$, is calculated using

$$\bar{x} = \frac{\sum_{i=1}^{n} f_i x_i}{\sum_{i=1}^{n} f_i},$$

where f_i is the number of values in the ith bin and x_i is the value of x corresponding to the mid-point of the ith bin. (c) A Gaussian probability density distribution with mean 2.5810 and standard deviation 0.0630. (d) A probability density distribution of means of samples of size 4.

distributions have the asymmetrical feature of a steep rise from the origin to the peak followed by a relatively gentle fall.

Next, 60 samples of size $n = 4$ are drawn at random from the population of values in list 10.1.[10] For each sample the four component values are given in list 10.2 at the end of this chapter (each with a number showing its location in list 10.1).

---

[10] We could choose a larger number of samples, but 60 samples, each of size 4, are sufficient to show how a coverage interval that contains the mean with high probability is obtained.

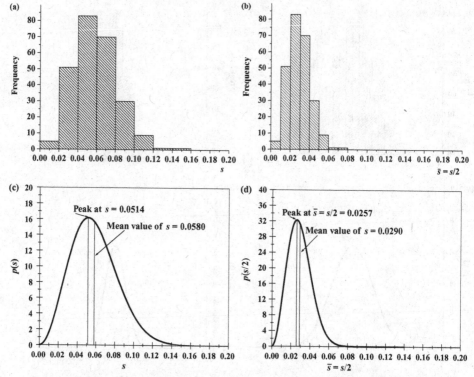

Figure 10.3. (a) A histogram of standard deviations of 250 samples of size 4. The mean of the histogram is 0.057 95. (b) A histogram of standard deviations of means of 250 samples of size 4. The mean of the histogram is 0.028 97. (c) The probability density distribution for sample standard deviation, s, for three degrees of freedom. The population standard deviation is 0.0630. (d) The probability density distribution for the standard deviation of means of sample sizes, $n = 4$. The population standard deviation is 0.0630.

The mean and standard deviation of the mean for each sample are also stated. The '95% coverage interval' for the population mean is then calculated on the evidence of each sample, using equation (10.3). We might anticipate that the probability that this interval encloses the population mean is 0.95 or 95%. For each of the 60 samples, this coverage interval is stated, and also whether or not this interval actually does enclose the population mean.

If we claim that each coverage interval has a probability of 95% of enclosing the population mean, and if we make 60 attempts at finding such a coverage interval, then the expected number of occasions when the true mean is actually in the interval should be $(95/100) \times 60$ or about 57. But, as indicated in list 10.2, the number of occasions when the population mean is enclosed within the coverage interval is only 52, which is about 87% of 60. It appears that the factor 1.96 in equation (10.3) should actually be somewhat larger.

If, instead of using s as the approximate unbiased estimate of σ, we used $1.086s$ as the exact unbiased estimate[11] of σ, we would have increased our success rate from 87% to only about 88%. Our failure to match expected and actual enclosure probabilities is not due to the use of an approximate unbiased estimate of s.

The explanation for the relatively low success rate in enclosing μ is that not only does $\bar{x}$ in equation (10.3) vary with the sample, *but so does s*. For three degrees of freedom as in this case, the variation of s is substantial and is shown in figure 10.3(a) for our particular Monte Carlo-derived population. Figures 10.3(b) and 10.3(d) show, respectively, the observed and theoretical variations in $s/\sqrt{n} = s/\sqrt{4} = s/2$ in our example where $n = 4$. The factor 1.96 in equation (10.3) entails the assumption that $s/\sqrt{n}$ is the *constant* standard deviation of $\bar{x}$ and that only $\bar{x}$ varies; the variation of $\bar{x}$ (a Gaussian variable) can then, on this assumption, be correctly described as covering the range $\pm 1.96s/\sqrt{n}$ for 95% of the time. Such a variation in $\bar{x}$ was illustrated in figure 10.1. But if s, and therefore $s/\sqrt{n}$, varies with the sample, the factor 1.96 cannot be correct for a 95% success rate, even though $\bar{x}$ remains a Gaussian variable. As we have just discovered, 1.96 must be replaced by a larger factor. On the other hand, for a larger number of degrees of freedom, as was shown in figure 9.3, the curve of s is narrower and so s is more nearly constant; 1.96 will then be closer to the correct factor for 95% coverage.

10.2 The coverage interval using a t-distribution

When the number of degrees of freedom is small, how do we find the factor that should replace 1.96 for 95% coverage? We note that, since equation (10.3) may be rewritten

$$\pm \frac{\bar{x} - \mu}{s/\sqrt{n}} = \text{a multiplying factor},$$

where the 'multiplying factor' is 1.96 for a 95% coverage interval and very many degrees of freedom (that is, the Gaussian situation), a promising approach for few degrees of freedom would be to regard the left-hand side, $(\bar{x} - \mu)/(s/\sqrt{n})$, as a new variable and to find its distribution. This new variable is called t_v and has a distribution called the t-distribution with v degrees of freedom.[12]

t_v is given by

$$t_v = \frac{\bar{x} - \mu}{s/\sqrt{n}}. \tag{10.4}$$

---

[11] See footnote 3 in this chapter.
[12] It is also known as 'Student's t', after the pen-name of W. S. Gosset, who published it in 1906.

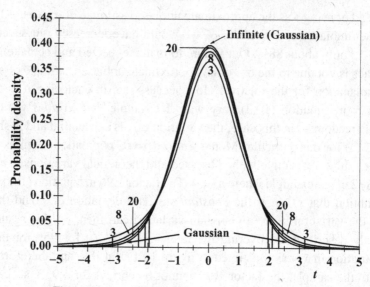

Figure 10.4. t-distributions for $\nu = 3, 8, 20$ and ∞.

Equation (10.4) can be written

$$\bar{x} = \mu \pm t_{X\%,\nu} \frac{s}{\sqrt{n}}, \tag{10.5}$$

where $t_{X\%,\nu}$ refers to the $X\%$ level of confidence for ν degrees of freedom. For very large ν and $X\% = 95\%$, $t_{X\%,\nu} = 1.96$. Conventionally, in deriving the mathematical formula for the t-distribution, μ is regarded as the fixed population parameter, and $\bar{x}$ and s as the variables that vary with the particular sample. The probability density, $p(t, \nu)$, of the t-distribution for ν degrees of freedom is given by[13]

$$p(t, \nu) = K(\nu)\left(1 + \frac{t^2}{\nu}\right)^{-(\nu+1)/2}, \tag{10.6}$$

where $K(\nu)$ ensures that the area under the probability density curve is unity.[14]

In equation (10.4), t_ν may be regarded as the difference between $\bar{x}$ and μ expressed in terms of the number of standard deviations of the mean, $s/\sqrt{n}$. We note that t_ν is a *dimensionless number*.

Figure 10.4 shows the probability density of the t-distribution for numbers of degrees of freedom $\nu = 3, 8, 20$ and ∞. The t-distribution is symmetric, even

---

[13] See Kendall and Stuart (1969).
[14] It may be shown that

$$K(\nu) = \frac{\Gamma\{(\nu + 1)/2\}}{\Gamma(\nu/2)}\sqrt{1/\pi\nu},$$

where Γ denotes the gamma function.

Table 10.1. *t values for ν degrees*
of freedom at the 95% level
of confidence

ν	$t_{95\%,\nu}$
3	3.18
8	2.31
20	2.09
∞	1.96

though it is the ratio of a Gaussian and therefore symmetrical distribution (the distribution of $\bar{x} - \mu$) to an asymmetrical distribution (the distribution of $s/\sqrt{n}$, as in figure 10.3(d)). For infinite ν, the *t*-distribution coincides exactly with the Gaussian distribution with mean zero and standard deviation 1. Figure 10.4 also shows the respective limits of the intervals along the horizontal axis which enclose 95% of the total area. For the Gaussian case (ν infinite), the limits are ±1.96. As ν decreases, the peak of the *t*-distribution is reduced and more of the area under the probability density curve is located in the tails.[15] As a consequence, as ν decreases, 95% of the total area is delimited by points further from the origin (which is at the centre of the horizontal axis). The limits for all four cases are given in table 10.1. Appendix A contains a more extensive table giving $t_{95\%,\nu}$ for a range of ν.

For samples of size $n = 4$ ($\nu = n - 1 = 3$), table 10.1 indicates that 3.18 should be used instead of 1.96 as the multiplier of the standard deviation of the mean. When this is done, the proportion of successful intervals – those enclosing the population mean – in list 10.2 increases to 56 out of 60, that is 93% of the intervals. This is much closer to the claimed 95% level of confidence, although we note that there is still statistical variability arising from the low number of 60 trials; a similar MCS with a much larger population and number of trials would have given a proportion of successful intervals much closer to 95%.

10.2.1 The coverage factor, k, and expanded uncertainty, U

The symbol $t_{X\%,\nu}$ in equation (10.5) is called the *coverage factor* and is given the more convenient symbol k. We therefore have the result that the standard uncertainty of an estimate multiplied by k gives the *expanded uncertainty*, U, of that estimate at that level of confidence (usually $X\% = 95\%$). Expanded uncertainty is given

---

[15] A lower peak must be accompanied by more area in the tails, since the total area beneath the curve must equal unity.

Table 10.2. *Variation of absorbance with concentration of standard silver solutions*

Concentration, C (ng/mL)	Absorbance, A (arbitrary units)
5.06	0.129
10.10	0.249
15.07	0.380
20.12	0.511
25.06	0.645

the upper-case symbol U, to distinguish it from standard uncertainty, u, so we have

$$U = ku. \tag{10.7}$$

It is conventional to quote an expanded uncertainty with a $\pm$ sign; for example, in an accurate measurement of length, U might be stated as $U = \pm 10$ μm. By contrast, a standard uncertainty should be stated without the $\pm$ symbol and indeed without any sign; thus u might be stated as $u = 5$ μm for that estimate of the measurand. It is uncommon for U to be quoted to more than two significant digits.

A generalised form of equation (10.4) can be used whenever a sample yields not just one least-squares estimate (the mean), but two or more. Two estimates might be the intercept, a, and slope, b, as when fitting the straight line $y = a + bx$ to x, y data. If the sample size is n, we now have $\nu = n - 2$ and, in place of equation (10.4), we have the following t-variables:

$$t_{X\%,\nu}^{(a)} = \frac{a - \alpha}{s_a}, \tag{10.8}$$

$$t_{X\%,\nu}^{(b)} = \frac{b - \beta}{s_b}. \tag{10.9}$$

Here a and b are unbiased estimates of the true intercept and slope, α and β, respectively. The standard uncertainties in a and b are s_a and s_b, respectively.[16]

Example 1

Equations (10.8) and (10.9) may be used to find coverage intervals. Table 10.2 contains data of absorbance, A (in arbitrary units), as a function of concentration, C, for standard silver solutions analysed by atomic absorption spectroscopy.

---

[16] We note that equations (10.8) and (10.9) do not have $1/\sqrt{n}$ in the denominator, whereas equation (10.4) does. However, in equation (10.4), $s/\sqrt{n}$ can be more briefly written $s_{\bar{x}}$, so all three equations are consistent in appearance when written in terms of the standard uncertainties of the *estimates* from the sample, namely $\bar{x}$ or a and b.

Table 10.3. *The area under an HPLC peak as a function of concentration*

Concentration (x) (mg/L)	Area (y) (arbitrary units)
1.006	8.20
2.012	17.6
5.030	42.8
7.555	65.7
10.064	90.5
15.101	136

Assuming the relationship between absorbance and concentration to be linear, we use least-squares to fit the equation

$$A = a + bC \qquad (10.10)$$

to the data in table 10.2, where a is the intercept and b is the slope.[17]

The least-squares estimate of the intercept is $a = -0.007\,35$, with standard uncertainty $s_a = 0.005\,49$. The least-squares estimate of the slope is $b = 0.025\,87$ mL/ng, with standard uncertainty $s_b = 0.000\,329$ mL/ng. Since there are five pairs of data in table 10.2 it follows that the number of degrees of freedom associated with the least-squares fit is $v = 5 - 2 = 3$. The t value for $v = 3$ is given in table 10.1 as $t_{95\%,3} = 3.18$. The expanded uncertainty, U, in a for the 95% level of confidence is $3.18 \times 0.005\,49 = \pm0.0175$. Similarly, the expanded uncertainty in b for the 95% level of confidence is $3.18 \times 0.000\,329$ mL/ng $= \pm0.00105$ mL/ng. We can now write

$a = -0.007 \pm 0.018$ and
$b = (0.0259 \pm 0.0011)$ mL/ng.

Exercise A
(1) An HPLC instrument was calibrated using known concentrations of sodium nitrate. Table 10.3 contains values of the concentration and area under a peak produced by the instrument.
Use the data in table 10.3 to
(a) find the slope and intercept of the best straight line through the data;
(b) calculate the standard uncertainty in the slope and intercept;
(c) find the expanded uncertainty in the best estimate of slope and intercept at the 95% level of confidence; and

---

[17] Details of fitting by least-squares are given in section 5.2.3.

(d) find the coverage intervals containing the true value of the slope and intercept at the 95% level of confidence.

(2) Using the data in table 5.2, find the coverage interval containing the true drift of the voltage reference at the 95% level of confidence.

10.3 The Welch–Satterthwaite formula

When inputs $x_1, x_2, \ldots, x_n$ are used to determine the best estimate of the measurand, y, through the functional relationship $y = f(x_1, x_2, \ldots, x_n)$, the combined standard uncertainty, $u(y)$, in y may be found using[18]

$$u^2(y) = c_1^2 u^2(x_1) + c_2^2 u^2(x_2) + \cdots + c_n^2 u^2(x_n), \tag{10.11}$$

where the c's are sensitivity coefficients defined by the partial derivatives, $c_i = \partial y / \partial x_i$ $(i = 1, 2, \ldots, n)$.

Each of the standard uncertainties, $u(x_i)$, of the inputs, x_i, is associated with ν_i degrees of freedom. If, for example, x_1 is the mean of ten repeated uncorrelated values that have a standard deviation s_1, then $u(x_1) = s_1 / \sqrt{10}$ has $\nu_1 = 9$ degrees of freedom.

The obvious question now is as follows: how many degrees of freedom should we associate with $u(y)$ on the left-hand side of equation (10.11)? The answer is provided by the Welch–Satterthwaite formula which, though only approximate, is nevertheless adequate for most cases.[19]

Consider two uncorrelated inputs, x_1 and x_2. In this case we have $y = f(x_1, x_2)$ and equation (10.11) may be written

$$u^2(y) = c_1^2 u^2(x_1) + c_2^2 u^2(x_2). \tag{10.12}$$

Let $u(x_1)$ and $u(x_2)$ be associated with ν_1 and ν_2 degrees of freedom, respectively. We now take the *variance* of both sides of equation (10.12). We recall that, for any constant, K, and variable x, $u^2(Kx) = K^2 u^2(x)$. Then

$$u^2[u^2(y)] = c_1^4 u^2[u^2(x_1)] + c_2^4 u^2[u^2(x_2)]. \tag{10.13}$$

We note another assumption: not only the inputs, x_1 and x_2, but also their variances, $u^2(x_1)$ and $u^2(x_2)$, are assumed to be uncorrelated. If the variances were correlated, equation (10.12) would contain a third term involving the covariance of the variances, $u^2(x_1)$ and $u^2(x_2)$.

Next we assume that the inputs, x_1 and x_2, are random Gaussian variables. As a consequence of the central limit theorem this assumption is likely to be valid

---

[18] This applies to uncorrelated inputs: see section 7.1.
[19] For further information see Ballico (2000) and Hall and Willink (2001).

if each of x_1 and x_2 is the mean of several values, and the greater the number of values, the better the approximation.[20] The central limit theorem allows a Gaussian distribution to be assumed as an approximation to the distribution of the *means* of randomly drawn samples, even if these samples are drawn from a non-Gaussian distribution.

An input, x_1, and its associated standard uncertainty, $u(x_1)$, may also be obtained from a calibration report or look-up table. To establish the standard uncertainty in the report, repeat measurements are likely to have been made. There is no difference in principle between a 'present' run that acquires several values through repeat measurements and a 'past' run; indeed, an uncertainty obtained through repeat measurements and classified as a Type A uncertainty (because of the statistical techniques involved in estimating it) is 'fossilised' into a Type B uncertainty when used subsequently. As a consequence, we may assume that a value, x_1, obtained from a calibration report or look-up table has a Gaussian distribution even though the associated standard uncertainty, $u(x_1)$, is Type B. Such an assumption also applies to the other input, x_2.

Calibration reports always state the uncertainty of a reported value, and sometimes also state the associated number of degrees of freedom. By contrast, look-up tables of properties of materials often give no indication of the uncertainty of the value of the quantity being looked up. The number of significant decimal places quoted can, however, be used to infer a rough figure for the uncertainty (see section 2.3). Because this inferred figure is only rough, estimated to perhaps no better than 30%, the associated number of degrees of freedom is low[21] (about six for 30% uncertainty). In all cases the uncertainty, whether explicitly stated or inferred, must refer to possible values of a quantity consistent with a distribution that has low-probability tails and a high-probability peak region. A Gaussian distribution best describes this situation.

In some situations we may need to determine an intercept and a slope from x, y data. Just as a mean will have a near-Gaussian distribution even when its component readings are drawn from a non-Gaussian distribution, so the intercept and slope will similarly have a near-Gaussian distribution. The reason is that the intercept and slope are calculated as a linear combination of the observed response variables, where the response variables are the y_i and the explanatory (error-free) variables are the x_i. It is this linear combination of possibly non-Gaussian variables that produces a near-Gaussian variable (and the larger the sample of such non-Gaussian variables, the closer will be the approximation to a Gaussian distribution).

---

[20] See, for example, section 8.6.
[21] Numbers of degrees of freedom are estimated if we can assess the uncertainty attaching to the uncertainty itself, as described by equation (9.18).

The above discussion suggests that, in most cases, we may take x_1 and x_2 in equation (10.13) as each having a Gaussian distribution. That being so, we now apply equation (9.7), repeated here:

$$u^2(s^2) = \frac{2\sigma^4}{\nu}. \tag{10.14}$$

We recall the meaning of equation (10.14): s^2 is the variance of a sample drawn from a Gaussian distribution with variance σ^2. Equation (10.14) gives the variance, $u^2(s^2)$, of s^2. This variance is based on ν degrees of freedom (for example, if a mean is calculated from n readings, then $\nu = n - 1$). The square root $u(s^2)$ of equation (10.14) is a measure of the 'fatness' of the curves in figure 9.2.

The term s^2 in equation (10.14) is equivalent to $u^2(x_1)$ or $u^2(x_2)$ in equation (10.13).[22] So we may write equation (10.13) as

$$u^2[u^2(y)] = \frac{2c_1^4\sigma_1^4}{\nu_1} + \frac{2c_2^4\sigma_2^4}{\nu_2}, \tag{10.15}$$

where σ_1^2 and σ_2^2 are the population variances of x_1 and x_2, respectively.

σ_1^2 is the same as $u^2(x_1)$, and σ_2^2 is the same as $u^2(x_2)$. Equation (10.15) may therefore be written

$$u^2[u^2(y)] = \frac{2c_1^4u^4(x_1)}{\nu_1} + \frac{2c_2^4u^4(x_2)}{\nu_2}. \tag{10.16}$$

We now claim that y has a near-Gaussian distribution. This is plausible for the following reason. Using c's for the sensitivity coefficients,

$$\delta y = c_1\,\delta x_1 + c_2\,\delta x_2. \tag{10.17}$$

The increments in equation (10.17) may be written $\delta y = y - \mu_y$, $\delta x_1 = x_1 - \mu_{x1}$ and $\delta x_2 = x_2 - \mu_{x2}$. The quantities μ_y, μ_{x1} and μ_{x2} are, respectively, the population means of y, x_1 and x_2. Thus, although the functional relationship, $y = f(x_1, x_2)$, may be highly nonlinear, small changes of y from its mean obey a linear relationship to small changes of x_1 and x_2 from their respective means. These changes are near-Gaussian (this is another interpretation of the statement that x_1 and x_2 are near-Gaussian), and therefore so is y.

If y is Gaussian, then we may assign an 'effective number of degrees of freedom', ν_{eff}, to $u^2(y)$; this is the purpose of the Welch–Satterthwaite formula, and equations (10.14) and (10.16) yield

$$u^2[u^2(y)] = \frac{2u^4(y)}{\nu_{\text{eff}}} = \frac{2c_1^4u^4(x_1)}{\nu_1} + \frac{2c_2^4u^4(x_2)}{\nu_2}. \tag{10.18}$$

---

[22] We have $u^2[u^2(x_1)] = 2\sigma_1^4/\nu_1$ and $u^2[u^2(x_2)] = 2\sigma_2^4/\nu_2$.

Since, from equation (10.11), $u^2(y) = c_1^2 u^2(x_1) + c_2^2 u^2(x_2)$, equation (10.18) gives, upon cancelling out the 2's,

$$\frac{\left[c_1^2 u^2(x_1) + c_2^2 u^2(x_2)\right]^2}{\nu_{\text{eff}}} = \frac{c_1^4 u^4(x_1)}{\nu_1} + \frac{c_2^4 u^4(x_2)}{\nu_2}. \tag{10.19}$$

Equation (10.19) may be rearranged as follows:

$$\nu_{\text{eff}} = \frac{\left[c_1^2 u^2(x_1) + c_2^2 u^2(x_2)\right]^2}{\dfrac{c_1^4 u^4(x_1)}{\nu_1} + \dfrac{c_2^4 u^4(x_2)}{\nu_2}}. \tag{10.20}$$

The effective number of degrees of freedom, ν_{eff}, is not necessarily an integer. In practice, ν_{eff} is often truncated to an integer for the purpose of calculating a coverage factor, k (for example, the numbers 6.2 and 6.8 would both truncate to 6). For x_i inputs, where $i = 1$ to n, equation (10.19) may be written generally as

$$\frac{\left[c_1^2 u^2(x_1) + c_2^2 u^2(x_2) + \cdots + c_n^2 u^2(x_n)\right]^2}{\nu_{\text{eff}}} = \frac{c_1^4 u^4(x_1)}{\nu_1} + \frac{c_2^4 u^4(x_2)}{\nu_2} + \cdots + \frac{c_n^4 u^4(x_n)}{\nu_n}. \tag{10.21}$$

Since the numerator on the left-hand side of equation (10.21) is $u^4(y)$, equation (10.21) may be written

$$\nu_{\text{eff}} = \frac{u^4(y)}{\displaystyle\sum_{i=1}^{n} \frac{c_i^4 u^4(x_i)}{\nu_i}}. \tag{10.22}$$

Equations (10.21) and (10.22) are equivalent statements of the Welch–Satterthwaite formula.

With ν_{eff} determined for $u(y)$ by equation (10.22), we can now regard the ratio $(y - \mu_y)/u(y)$ as a t-variable for ν_{eff} degrees of freedom:

$$t_{\nu_{\text{eff}}} = \frac{y - \mu_y}{u(y)}. \tag{10.23}$$

Equation (10.23) is analogous to, and should be compared with, equations (10.4), (10.8) and (10.9). Coverage intervals for μ_y are now obtainable in the manner described in section 10.2. If, for example, $\nu_{\text{eff}} = 8$, the 95% coverage interval for μ_y is $\mu_y \pm 2.31 \times u(y)$, in which μ_y is estimated by y as obtained from the inputs $x_1, x_2, \ldots, x_n$ and $u(y)$ is given by equation (10.11). The expanded uncertainty $U(y)$ is given, in this case, by $U(y) = 2.31 u(y)$.

The determination of the expanded uncertainty, $U(y)$, in the measurand, y, represents the conclusion of the process of measuring y. For this process we need to know the values of the n inputs, $x_1, x_2, \ldots, x_n$, their standard uncertainties $u(x_1)$, $u(x_2), \ldots, u(x_n)$, their associated degrees of freedom $\nu_1, \nu_2, \ldots, \nu_n$ and the sensitivity coefficients (the partial derivatives) $c_1, c_2, \ldots, c_n$. When there are many inputs, the calculations can be lengthy and are often neatly summarised by means of a table (sometimes referred to as an 'uncertainty budget'). Practical advice for such cases is available.[23]

If, when we have two inputs, $c_1 = c_2 = 1$ (sensitivity coefficients are in fact often equal to 1) and also $u(x_1) = u(x_2)$ ($u(x_1)$ and $u(x_2)$ being mutually independent), equation (10.20) then gives

$$\nu_{\text{eff}} = \frac{4\nu_1\nu_2}{\nu_1 + \nu_2}. \tag{10.24}$$

Equation (10.24) implies that, if, for example, a Type A uncertainty, $u(x_1)$, with a large number of degrees of freedom, ν_1, is combined with a roughly similar Type B uncertainty, $u(x_2)$, which has a small number of degrees of freedom, ν_2, then the combined uncertainty $\sqrt{u^2(x_1) + u^2(x_2)}$ will have an associated ν_{eff} closer to the lower of ν_1 and ν_2. If $\nu_1 = 100$ and $\nu_2 = 5$, then (using equation (10.24) $\nu_{\text{eff}} \simeq 19$.

We note from equation (10.21) that a high $c_k u(x_k)$, for any particular input x_k, and a low associated ν_k reinforce each other to make that kth term dominant on the right-hand side of equation (10.21). A high $c_k u(x_k)$ or a low ν_k, or both, are often the effect of a systematic error. Then we have $\nu_{\text{eff}} \simeq \nu_k$, so the uncertainty of the measurand is dominated by the least accurate input and has a number of degrees of freedom not much different from the low number of degrees of freedom of that input. Owing to the likely presence of systematic errors the accuracy of a measurement cannot be significantly improved, beyond a certain point, merely by increasing the number of readings. A high value of uncertainty, $u(x_k)$, however, might not be important if the measurand is insensitive to the value of that input (small c_k).

An instructive case of equation (10.21) occurs when y is the mean of repeated readings $x_1, x_2, \ldots, x_n$:

$$y = \frac{x_1 + x_2 + \cdots + x_n}{n}. \tag{10.25}$$

We assume that all the x_i ($i = 1, 2, \ldots, n$) are independently drawn from one population (since they are independently drawn, they are uncorrelated). The variance of all the x_i has the same value, $u^2(x)$, as the variance of the population from which they were sampled. Equation (10.25) with uncorrelated x_i implies that

$$u^2(y) = \frac{u^2(x)}{n}. \tag{10.26}$$

---

[23] See, for example, Bentley (2005).

With $\nu_x = n - 1$ as the number of degrees of freedom associated with $u^2(x)$, equations (10.26) and (10.14) give

$$u^2[u^2(y)] = \frac{1}{n^2}u^2[u^2(x)] = \frac{1}{n^2}\frac{2u^4(x)}{n-1},\tag{10.27}$$

so that, setting $u^2[u^2(y)] = 2u^4(y)/\nu_{\text{eff}}$ as before, equation (10.27) gives

$$\frac{2u^4(y)}{\nu_{\text{eff}}} = \frac{2}{n^2}\frac{u^4(x)}{n-1},\tag{10.28}$$

and since, from equation (10.26), $u^4(y) = u^4(x)/n^2$, equation (10.28) gives finally

$$\nu_{\text{eff}} = n - 1.\tag{10.29}$$

The variance, $u^2(y)$, and therefore the standard deviation or standard uncertainty, $u(y)$, of y, are associated with the same number of degrees of freedom as $u^2(x)$, which is the (unbiased) variance of the n values of repeated readings x_i ($i = 1, 2, \ldots, n$). This result is to be expected: because $u^2(y) = u^2(x)/n$, the sampling distribution of $u^2(y)$ must be a scaled version of the sampling distribution of $u^2(x)$ for the particular number of degrees of freedom (as shown in figure 9.2 for several values of ν). This scaled version for $u^2(y)$ must keep the same shape as for $u^2(x)$, so the number of degrees of freedom associated with $u^2(y)$ must also be the same.

In the demonstration of the Welch–Satterthwaite formula, we made the assumption that not only are the x_i uncorrelated, but so also are the $u^2(x_i)$. However, for this particular case $y = (x_1 + x_2 + \cdots + x_n)/n$, the fact that all the x_i are drawn from the same population with variance $u^2(x)$ implies that the $u^2(x_i)$ are *not* uncorrelated; in fact, we now have all the $u^2(x_i)$ equal at the value $u^2(x)$, and therefore perfectly correlated! Although the result $\nu_{\text{eff}} = n - 1$ in equation (10.29) is correct and was shown using $u^2(y) = u^2(x)/n$, a full demonstration from first principles starting from equation (10.12) (generalised to n inputs) would need to take into account the correlation between the variances. The step from equation (10.12) to equation (10.13) would now be invalid; equation (10.13) would have additional terms corresponding to the correlation terms in equation (7.36), and it can be shown that equation (10.13) with the necessary additional terms leads to the same result $\nu_{\text{eff}} = n - 1$.

In calculations involving the Welch–Satterthwaite formula, it is prudent to keep extra decimal places when evaluating standard uncertainties. This is a consequence of the fourth powers in the formula, which may easily create round-off errors in the final result for the effective number of degrees of freedom, and hence in the coverage interval.

Example 2

The moment of inertia, I, of a solid cylinder of mass M, rotating about its principal axis, is given by[24]

$$I = \frac{MR^2}{2},$$
(10.30)

where R is the radius of the cylinder. The mean of eight values of the mass measured in an experiment is 252.6 g and the standard uncertainty in the mean mass is 2.5 g. The mean of five values of the radius is 6.35 cm with a standard uncertainty in the mean radius of 0.05 cm. Use this information to determine

(a) the best estimate for the moment of inertia of the cylinder;
(b) the standard uncertainty in the best estimate of the moment of inertia, assuming that the errors in the mass and radius measurements are uncorrelated;
(c) the effective number of degrees of freedom of the measurand uncertainty using the Welch–Satterthwaite formula;
(d) the coverage factor for the 95% level of confidence; and
(e) the coverage interval containing I at the 95% level of confidence.

Answer

(a) Using equation (10.30), the best estimate of the moment of inertia is

$$I = \frac{252.6 \times (6.35)^2}{2} = 5092.7 \text{ g} \cdot \text{cm}^2.$$

(b) Following equation (10.12), can write the variance in the best estimate as

$$u^2(I) = c_M^2 u^2(M) + c_R^2 u^2(R),$$
(10.31)

where

$$c_M = \frac{\partial I}{\partial M} = \frac{R^2}{2}, \qquad c_R = \frac{\partial I}{\partial R} = MR.$$

Equation (10.31) becomes

$$u^2(I) = \left(\frac{R^2}{2}\right)^2 u^2(M) + (MR)^2 u^2(R)$$

$$= \left(\frac{(6.35)^2}{2}\right)^2 \times (2.5)^2 + (252.6 \times 6.35)^2 \times (0.05)^2$$

$$= 2540.5 + 6432.1 = 8972.6 \, (\text{g} \cdot \text{cm}^2)^2.$$

It follows that $u(I) = 94.7 \text{g} \cdot \text{cm}^2$.

---

[24] See Young and Freedman (2003).

(c) Replacing the subscripts in equation (10.20) by M and R as appropriate, we can write the Welch–Satterthwaite formula for this example as

$$\nu_{\text{eff}} = \frac{\left[c_M^2 u^2(M) + c_R^2 u^2(R)\right]^2}{\dfrac{c_M^4 u^4(M)}{\nu_M} + \dfrac{c_R^4 u^4(R)}{\nu_R}},$$

where ν_M is the number of degrees of freedom in the calculation of the standard uncertainty in M, i.e. $\nu_M = 8 - 1 = 7$. The number of degrees of freedom ν_R in the calculation of the standard uncertainty in R is $\nu_R = 5 - 1 = 4$. From part (b), $c_M^2 u^2(M) = 2540.5$ and $c_R^2 u^2(R) = 6432.1$, so

$$\nu_{\text{eff}} = \frac{(2540.5 + 6432.1)^2}{\dfrac{(2540.5)^2}{7} + \dfrac{(6432.1)^2}{4}} = 7.1,$$

which truncates to 7 for the purpose of calculating the coverage factor, k.

(d) The t value for the 95% level of confidence and seven degrees of freedom is found from the table in appendix B. We have $k = t_{95\%,7} = 2.36$.

(e) The interval containing the true value at the 95% level of confidence is $5092.7 \text{ g} \cdot \text{cm}^2 \pm 2.36 \times 94.7 \text{ g} \cdot \text{cm}^2 = (5092.7 \pm 223.5) \text{ g} \cdot \text{cm}^2$. The moment of inertia may be expressed in scientific notation to an appropriate number of significant figures as

$$\text{Moment of inertia of the cylinder} = (5.09 \pm 0.22) \times 10^3 \text{g} \cdot \text{cm}^2.$$

Example 3

In this example we include the influence of resolution when calculating a confidence interval.

The contribution to the combined standard uncertainty due to the resolution of an instrument is worthy of special mention. Resolution is a perennial (though often small) contributor to the combined standard uncertainty. The manner by which this contribution is quantified is still a matter of research and debate.[25] Here we have adopted the approach suggested by the GUM.[26]

The diameter of a wire is measured five times using a micrometer with a resolution of 0.01 mm. The mean diameter is found to be 0.253 mm with a standard uncertainty in the mean of 0.007 mm. Use this information to calculate

(a) the best estimate of the cross-sectional area of the wire;
(b) the standard uncertainty in the best estimate;
(c) the effective number of degrees of freedom for the standard uncertainty;
(d) the coverage factor, k, for the 95% level of confidence; and

---

[25] See, for example, Elster (2000) and Frenkel and Kirkup (2005).
[26] This is consistent with advice contained in the GUM (see annex F to the GUM (1995)); see also Lira and Wöger (1997).

(e) the coverage interval containing the true value of the cross-sectional area of the wire at the 95% level of confidence.

Answer

(a) The value of the cross-sectional area, A, of the wire is given by

$$A = \frac{\pi D^2}{4},$$
(10.32)

where D is diameter of the wire. D may be written as

$$D = X + Z.$$
(10.33)

X is the mean diameter of the wire obtained by calculating the mean of repeat values of the diameter. Z is the correction required due to systematic errors. From the information in this example, $X = 0.253$ mm. Since the correction term due to the resolution of the instrument is as likely to be positive as negative, we take $Z = 0$. It follows that $D = 0.253$ mm $+ 0 = 0.253$ mm.

Substituting $D = 0.253$ mm into equation (10.32) gives $A = 0.0503$ mm^2.

(b) The standard uncertainty in the diameter, $u(D)$, can be found using

$$u^2(D) = u^2(X) + u^2(Z).$$
(10.34)

$u(X)$ is given in the question as equal to 0.007 mm. $u(Z)$ is determined by assuming that the probability distribution associated with the scatter of Z is rectangular with a width of $\delta = 0.01$ mm, in which case the standard deviation is $\delta/\sqrt{12}$ (see section 8.3). Then $u(Z_D) = 0.01$ mm$/\sqrt{12} = 2.9 \times 10^{-3}$ mm. Using equation (10.34), we obtain

$$u^2(D) = (7 \times 10^{-3})^2 + (2.9 \times 10^{-3})^2$$
$$= 4.9 \times 10^{-5} + 8.33 \times 10^{-6}$$
$$= 5.73 \times 10^{-5} \, \text{mm}^2.$$

It follows that $u(D) = 7.6 \times 10^{-3}$ mm.

(c) Following equation (10.20), we write ν_{eff} as

$$\nu_{\text{eff}} = \frac{\left[c_X^2 u^2(X) + c_Z^2 u^2(Z)\right]^2}{\dfrac{c_X^4 u^4(X)}{\nu_X} + \dfrac{c_Z^4 u^4(Z)}{\nu_Z}}.$$
(10.35)

Now, using equation (10.33),

$$c_X = \frac{\partial D}{\partial X} = 1$$

and

$$c_Z = \frac{\partial D}{\partial Z} = 1.$$

Therefore equation (10.35) simplifies to

$$v_{\text{eff}} = \frac{[u^2(X) + u^2(Z)]^2}{\dfrac{u^4(X)}{v_X} + \dfrac{u^4(Z)}{v_Z}}. \tag{10.36}$$

Now $v_X = 5 - 1 = 4$. Since the uncertainty in the standard uncertainty in Z is zero, equation (10.14) indicates that the effective number of degrees of freedom is very large: v_Z tends to ∞. Equation (10.36) becomes

$$v_{\text{eff}} = \frac{(5.73 \times 10^{-5})^2}{\dfrac{(7 \times 10^{-3})^4}{4} + 0} = 5.5,$$

which truncates to $v_{\text{eff}} = 5$.

(d) The t value for the 95% confidence interval for D based on five degrees of freedom is found from the table in Appendix B to be $t_{95\%,5} = 2.57$, so that $k = 2.57$.

(e) To calculate the coverage interval containing the true value of the area at the 95% level of confidence, we write

$$u(A) = \left(\frac{\partial A}{\partial D}\right) u(D).$$

From equation (10.32)

$$\frac{\partial A}{\partial D} = \frac{\pi D}{2} = \frac{\pi \times 0.253}{2} = 0.397 \text{ mm},$$

so

$$u(A) = 0.397 \times 7.6 \times 10^{-3} = 0.0030 \text{ mm}^2.$$

It follows that the coverage interval containing the true value of the cross-sectional area at the 95% level of confidence is $(0.0503 \pm 2.571 \times 0.0030)$ mm^2, i.e.

$$\text{Cross-sectional area} = (0.0503 \pm 0.0077) \text{ mm}^2.$$

Exercise B

The volume, V, of a cylinder of length, L, and radius, r, is given by

$$V = \pi r^2 L.$$

Four measurements of the length of the cylinder and five of its radius are made. The mean length is 15.3 cm with a standard uncertainty in the mean length of 0.1 cm. The mean radius is 3.85 cm with a standard uncertainty in the mean radius of 0.02 cm. Use this information to determine

(a) the best estimate for the volume of the cylinder;
(b) the standard uncertainty in the best estimate of the cylinder's volume, assuming that the errors in the mass and radius measurements are uncorrelated;

(c) the effective number of degrees of freedom of the measurand uncertainty using the Welch–Satterthwaite formula;
(d) the coverage factor for the 95% level of confidence; and
(e) the coverage interval containing the true value of V at the 95% level of confidence.

Exercise C

n independent readings are obtained using a DMM. The standard deviation of the values is 30 μV (microvolts). The DMM has a systematic error of −50 μV for the particular range of values being measured. (Each reading is therefore increased by 50 μV). This systematic error has an estimated standard uncertainty of 10 μV on six degrees of freedom. Consider two situations.

(1) Presence of systematic error.
   (a) If ten readings are taken, and their mean calculated, find (i) the resultant standard uncertainty of the mean reading; (ii) its effective number of degrees of freedom; and (iii) the expanded uncertainty of the mean reading for a 95% coverage interval.
   (b) Suppose that, in (a), the number of readings is doubled to 20. What are now the values of (i), (ii) and (iii) above?
(2) Absence of systematic error.
   (a) Repeat (1) (a) (ten readings), assuming that the DMM has *no* systematic error.
   (b) Repeat (1) (b) (20 readings), again assuming that the DMM has no systematic error.

10.3.1 The effective number of degrees of freedom v_{eff} can never exceed the sum of the numbers of degrees of freedom of the inputs

With n inputs $x_1, x_2, \ldots, x_n$ and standard uncertainties $u(x_1), u(x_2), \ldots, u(x_n)$ on $v_1, v_2, \ldots, v_n$ degrees of freedom, respectively, we always have

$$v_{\text{eff}} \le v_1 + v_2 + \cdots + v_n. \tag{10.37}$$

An equals sign would appear in equation (10.37) when the variance terms are in the same mutual ratio as the respective numbers of degrees of freedom:

$$\frac{c_1^2 u^2(x_1)}{c_2^2 u^2(x_2)} = \frac{v_1}{v_2},$$

$$\frac{c_1^2 u^2(x_1)}{c_3^2 u^2(x_3)} = \frac{v_1}{v_3},$$

and similarly for every pair of inputs. If all these conditions are satisfied, then $v_{\text{eff}} = v_1 + v_2 + \cdots + v_n$. This follows from Equation (10.21) and may be shown by dividing both sides of the equation by (for example) $c_1^4 u^4(x_1)$. These conditions are satisfied extremely rarely, if ever. So the *inequality* in Equation (10.37) is in practice always observed: the effective number of degrees of freedom associated with the uncertainty of the measurand is *less* than the sum of the individual numbers of degrees of freedom associated with the uncertainties of the inputs.[27]

Equation (10.37) may be verified by algebraic manipulation of equation (10.21). An alternative demonstration, using an electric analogue, is given in Appendix C.

10.4 Review

To determine the expanded uncertainty in a measurand, we need to know the effective number of degrees of freedom to be associated with the standard uncertainty in the measurand. This number of degrees of freedom is obtained using the Welch–Satterthwaite formula, for which we need to know in advance the standard uncertainties in the inputs to the measurand and the numbers of degrees of freedom associated with them. The resultant effective number of degrees of freedom gives us the coverage factor for a particular level of confidence, usually 95%. The coverage factor multiplied by the standard uncertainty in the measurand gives the expanded uncertainty in the measurand. In the next chapter we apply these methods to the calculation of uncertainties in a selection of typical experiments carried out in undergraduate laboratories.

---

[27] It would be very surprising if the uncertainty of the measurand had *more* degrees of freedom than the sum of the component degrees of freedom contributed by the uncertainties of the inputs. This would amount to a metrological 'free lunch'!

List 10.1. One thousand random numbers from a population with mean $\mu = 2.5810$ and standard deviation $\sigma = 0.0630$.

```
                        Gaussian population of 1000:
                        Mean m = 2.581, st. dev. 0.0630
                        Ident. no. of value and value:
    . 1 2.6821    2 2.3900    3 2.5799    4 2.5869    5 2.6350
      6 2.5632    7 2.5594    8 2.5381    9 2.5812   10 2.5425
     11 2.6418   12 2.5436   13 2.5671   14 2.6177   15 2.4331
     16 2.5077   17 2.5796   18 2.5569   19 2.6022   20 2.5095
     21 2.5200   22 2.5844   23 2.5861   24 2.6365   25 2.4893
     26 2.5255   27 2.5947   28 2.5340   29 2.4838   30 2.4852
     31 2.5021   32 2.7011   33 2.6461   34 2.5320   35 2.6743
     36 2.4549   37 2.5590   38 2.6885   39 2.6190   40 2.5962
     41 2.5860   42 2.5606   43 2.6159   44 2.6772   45 2.5967
     46 2.5655   47 2.6140   48 2.5716   49 2.4901   50 2.6875
     51 2.6106   52 2.5955   53 2.6203   54 2.5374   55 2.5762
     56 2.6098   57 2.4775   58 2.5357   59 2.5295   60 2.5603
     61 2.4647   62 2.4846   63 2.5436   64 2.5617   65 2.6462
     66 2.4977   67 2.5659   68 2.6266   69 2.5785   70 2.6658
     71 2.5840   72 2.4764   73 2.5560   74 2.6349   75 2.5454
     76 2.6350   77 2.5855   78 2.5727   79 2.6286   80 2.7078
     81 2.5448   82 2.5424   83 2.6305   84 2.6124   85 2.6819
     86 2.6728   87 2.5839   88 2.7174   89 2.5607   90 2.6169
     91 2.7514   92 2.6370   93 2.5783   94 2.5635   95 2.6203
     96 2.6520   97 2.5341   98 2.5561   99 2.6046  100 2.5590
    101 2.4925  102 2.5334  103 2.4762  104 2.5911  105 2.5305
    106 2.5781  107 2.5872  108 2.5235  109 2.4816  110 2.6072
    111 2.4775  112 2.6642  113 2.6463  114 2.6300  115 2.5891
    116 2.5014  117 2.5789  118 2.5488  119 2.5072  120 2.6146
    121 2.5828  122 2.4949  123 2.4353  124 2.5916  125 2.5586
    126 2.5178  127 2.5557  128 2.5975  129 2.6632  130 2.6044
    131 2.6117  132 2.6149  133 2.5138  134 2.5548  135 2.6706
    136 2.5982  137 2.5535  138 2.5772  139 2.7399  140 2.6061
    141 2.6811  142 2.6818  143 2.5805  144 2.5712  145 2.4313
    146 2.5612  147 2.6410  148 2.6269  149 2.6209  150 2.5561
    151 2.6480  152 2.6731  153 2.6065  154 2.6030  155 2.6387
    156 2.5882  157 2.6564  158 2.4720  159 2.6577  160 2.5207
    161 2.4979  162 2.5125  163 2.5410  164 2.6311  165 2.7157
    166 2.5774  167 2.5456  168 2.5671  169 2.6684  170 2.6231
    171 2.5676  172 2.6321  173 2.5791  174 2.5406  175 2.6719
    176 2.6175  177 2.6938  178 2.6618  179 2.6164  180 2.6033
    181 2.6622  182 2.5907  183 2.5760  184 2.5710  185 2.5737
    186 2.5734  187 2.5605  188 2.5704  189 2.4275  190 2.6142
    191 2.6142  192 2.6123  193 2.6273  194 2.5967  195 2.4492
    196 2.5296  197 2.5793  198 2.6356  199 2.5098  200 2.5344
    201 2.6815  202 2.5719  203 2.5922  204 2.4992  205 2.7342
    206 2.5777  207 2.6419  208 2.6534  209 2.5500  210 2.6053
    211 2.6711  212 2.6150  213 2.5955  214 2.5248  215 2.5902
    216 2.5028  217 2.4321  218 2.5558  219 2.6257  220 2.4339
    221 2.6461  222 2.5308  223 2.5865  224 2.6019  225 2.5931
    226 2.5861  227 2.6007  228 2.5308  229 2.5118  230 2.7241
    231 2.6405  232 2.4954  233 2.5254  234 2.5616  235 2.6028
    236 2.6731  237 2.5917  238 2.5321  239 2.5976  240 2.6646
    241 2.5422  242 2.5704  243 2.6210  244 2.6510  245 2.5793
    246 2.5862  247 2.5587  248 2.5310  249 2.5114  250 2.6188
    251 2.5383  252 2.6834  253 2.6682  254 2.5408  255 2.5657
    256 2.5221  257 2.5780  258 2.5968  259 2.5950  260 2.4376
    261 2.6685  262 2.5921  263 2.6278  264 2.6144  265 2.5884
    266 2.5887  267 2.7066  268 2.5990  269 2.5503  270 2.5976
    271 2.5737  272 2.5951  273 2.6044  274 2.5443  275 2.6434
    276 2.6371  277 2.6443  278 2.6042  279 2.4998  280 2.6088
    281 2.7315  282 2.6083  283 2.6782  284 2.6008  285 2.5948
    286 2.5528  287 2.5765  288 2.5152  289 2.5988  290 2.5775
    291 2.5982  292 2.6053  293 2.5312  294 2.5641  295 2.6302
    296 2.6908  297 2.7090  298 2.5476  299 2.6222  300 2.5641
    301 2.6138  302 2.5909  303 2.4896  304 2.5804  305 2.7151
    306 2.5511  307 2.5409  308 2.5848  309 2.6251  310 2.4644
    311 2.5573  312 2.6541  313 2.5297  314 2.6832  315 2.5730
    316 2.6705  317 2.5578  318 2.5717  319 2.5837  320 2.6258
    321 2.5884  322 2.6524  323 2.4404  324 2.5822  325 2.6417
    326 2.6810  327 2.5389  328 2.6256  329 2.5213  330 2.6358
    331 2.6195  332 2.4640  333 2.6040  334 2.5945  335 2.6566
    336 2.6154  337 2.6340  338 2.5184  339 2.5327  340 2.6178
    341 2.5736  342 2.4596  343 2.5678  344 2.5927  345 2.5493
    346 2.6370  347 2.5185  348 2.3962  349 2.6470  350 2.5417
    351 2.4555  352 2.6284  353 2.4618  354 2.4852  355 2.4707
    356 2.5525  357 2.5259  358 2.6215  359 2.6074  360 2.6056
    361 2.5320  362 2.6811  363 2.6311  364 2.5449  365 2.5830
    366 2.4986  367 2.5652  368 2.6025  369 2.5449  370 2.5914
    371 2.5832  372 2.7253  373 2.6008  374 2.5576  375 2.5312
    376 2.4845  377 2.5711  378 2.5847  379 2.5094  380 2.5227
    381 2.5754  382 2.5719  383 2.5290  384 2.6329  385 2.5589
    386 2.5790  387 2.5069  388 2.6049  389 2.4467  390 2.5123
    391 2.6010  392 2.5370  393 2.5767  394 2.6700  395 2.4789
    396 2.5263  397 2.5611  398 2.5437  399 2.6054  400 2.5392
    401 2.6297  402 2.6606  403 2.5201  404 2.5741  405 2.5905
    406 2.5357  407 2.5186  408 2.6264  409 2.4893  410 2.6313
    411 2.6708  412 2.5530  413 2.5989  414 2.6573  415 2.5520
    416 2.5201  417 2.6966  418 2.6247  419 2.5773  420 2.5237
    421 2.7304  422 2.6872  423 2.6167  424 2.6128  425 2.5609
```

```
426 2.5796 427 2.5918 428 2.5398 429 2.5588 430 2.5266
431 2.6410 432 2.5364 433 2.5431 434 2.6496 435 2.4648
436 2.5072 437 2.5414 438 2.5937 439 2.5470 440 2.6582
441 2.6992 442 2.6360 443 2.6123 444 2.5614 445 2.5381
446 2.5443 447 2.6526 448 2.5929 449 2.4805 450 2.4758
451 2.6286 452 2.4971 453 2.5545 454 2.5513 455 2.4842
456 2.6919 457 2.5907 458 2.6177 459 2.6451 460 2.6080
461 2.5370 462 2.5140 463 2.6078 464 2.4393 465 2.6373
466 2.5121 467 2.6125 468 2.5635 469 2.5221 470 2.5358
471 2.5679 472 2.5045 473 2.5087 474 2.4855 475 2.5862
476 2.4925 477 2.5642 478 2.4941 479 2.5578 480 2.5229
481 2.3987 482 2.5681 483 2.5268 484 2.5386 485 2.5985
486 2.5714 487 2.6493 488 2.6057 489 2.6352 490 2.4294
491 2.5798 492 2.5698 493 2.5282 494 2.4806 495 2.6605
496 2.5534 497 2.7012 498 2.5085 499 2.5311 500 2.6673
501 2.5847 502 2.6537 503 2.6385 504 2.7383 505 2.6508
506 2.6140 507 2.6012 508 2.5605 509 2.5722 510 2.5651
511 2.6427 512 2.5361 513 2.6555 514 2.6298 515 2.5153
516 2.6366 517 2.6781 518 2.5726 519 2.6140 520 2.6102
521 2.6314 522 2.4839 523 2.5767 524 2.6214 525 2.6239
526 2.5387 527 2.6258 528 2.6380 529 2.5434 530 2.5517
531 2.7154 532 2.5903 533 2.5213 534 2.5648 535 2.5880
536 2.5058 537 2.7162 538 2.6670 539 2.6689 540 2.5753
541 2.4893 542 2.6135 543 2.5975 544 2.6424 545 2.5529
546 2.5822 547 2.4953 548 2.7318 549 2.6604 550 2.5469
551 2.6082 552 2.6084 553 2.6504 554 2.5565 555 2.6334
556 2.5438 557 2.7288 558 2.5864 559 2.4015 560 2.4560
561 2.6384 562 2.6135 563 2.6468 564 2.6803 565 2.6680
566 2.5996 567 2.5893 568 2.5093 569 2.4508 570 2.6172
571 2.5626 572 2.6220 573 2.5737 574 2.6143 575 2.5641
576 2.5476 577 2.5986 578 2.6038 579 2.5964 580 2.6020
581 2.5472 582 2.5454 583 2.4921 584 2.5756 585 2.5801
586 2.5765 587 2.6056 588 2.4897 589 2.5790 590 2.5573
591 2.5672 592 2.5312 593 2.4606 594 2.6302 595 2.5851
596 2.5816 597 2.5699 598 2.6422 599 2.6393 600 2.5428
601 2.5776 602 2.4564 603 2.6395 604 2.6109 605 2.5906
606 2.5711 607 2.5363 608 2.5800 609 2.5646 610 2.5454
611 2.4427 612 2.5808 613 2.6295 614 2.7109 615 2.5507
616 2.4908 617 2.6103 618 2.5045 619 2.6471 620 2.6624
621 2.6332 622 2.5415 623 2.6193 624 2.6788 625 2.5744
626 2.6589 627 2.4806 628 2.5027 629 2.5712 630 2.5988
631 2.6426 632 2.5826 633 2.5366 634 2.4940 635 2.5952
636 2.6552 637 2.6031 638 2.5437 639 2.5920 640 2.5600
641 2.5216 642 2.4724 643 2.5854 644 2.5211 645 2.5752
646 2.6069 647 2.6192 648 2.5597 649 2.5259 650 2.5778
651 2.5409 652 2.6569 653 2.5527 654 2.6354 655 2.5007
656 2.6691 657 2.6354 658 2.5728 659 2.6103 660 2.5901
661 2.7015 662 2.7107 663 2.5340 664 2.5146 665 2.5068
666 2.6885 667 2.6034 668 2.6050 669 2.6342 670 2.6161
671 2.6030 672 2.5670 673 2.5124 674 2.5153 675 2.5327
676 2.6281 677 2.4555 678 2.6790 679 2.4399 680 2.5564
681 2.5661 682 2.5830 683 2.5821 684 2.5959 685 2.6962
686 2.5242 687 2.6240 688 2.6770 689 2.6218 690 2.5445
691 2.5323 692 2.5918 693 2.5076 694 2.5939 695 2.6192
696 2.6744 697 2.5274 698 2.5914 699 2.5488 700 2.6252
701 2.6044 702 2.5017 703 2.4788 704 2.5917 705 2.6271
706 2.6179 707 2.5950 708 2.5456 709 2.5699 710 2.6697
711 2.6349 712 2.5112 713 2.4589 714 2.5955 715 2.6376
716 2.6657 717 2.4924 718 2.6090 719 2.5204 720 2.5590
721 2.5922 722 2.5641 723 2.5781 724 2.6362 725 2.6357
726 2.6396 727 2.5225 728 2.7086 729 2.6176 730 2.5609
731 2.4638 732 2.5911 733 2.5673 734 2.5158 735 2.6330
736 2.5133 737 2.6396 738 2.4129 739 2.5473 740 2.4738
741 2.5440 742 2.5505 743 2.5478 744 2.6340 745 2.6273
746 2.5740 747 2.5967 748 2.7024 749 2.6492 750 2.5536
751 2.5991 752 2.6294 753 2.5498 754 2.4134 755 2.6708
756 2.6201 757 2.5913 758 2.4996 759 2.6710 760 2.6303
761 2.5281 762 2.6492 763 2.5366 764 2.5659 765 2.5139
766 2.4603 767 2.5702 768 2.5350 769 2.5810 770 2.5078
771 2.6212 772 2.5949 773 2.6212 774 2.6043 775 2.6130
776 2.5460 777 2.6157 778 2.6047 779 2.5400 780 2.5680
781 2.5989 782 2.5361 783 2.6271 784 2.6323 785 2.5219
786 2.5823 787 2.6447 788 2.5776 789 2.6109 790 2.5452
791 2.5419 792 2.6056 793 2.5594 794 2.5715 795 2.6275
796 2.5008 797 2.6224 798 2.4623 799 2.5995 800 2.6202
801 2.5954 802 2.5716 803 2.5608 804 2.5362 805 2.6429
806 2.5484 807 2.6487 808 2.5960 809 2.4771 810 2.5111
811 2.5169 812 2.6106 813 2.5310 814 2.5615 815 2.4915
816 2.5927 817 2.6440 818 2.5539 819 2.5923 820 2.5766
821 2.6206 822 2.5668 823 2.6033 824 2.5793 825 2.5125
826 2.6458 827 2.6152 828 2.6685 829 2.4878 830 2.6148
831 2.6277 832 2.5435 833 2.6471 834 2.6889 835 2.6019
836 2.6282 837 2.5029 838 2.6503 839 2.6002 840 2.4718
841 2.5272 842 2.6564 843 2.7571 844 2.6263 845 2.5712
846 2.4777 847 2.4818 848 2.6409 849 2.5439 850 2.5794
851 2.6555 852 2.5878 853 2.5836 854 2.5492 855 2.6077
856 2.6296 857 2.5770 858 2.6537 859 2.6374 860 2.5267
861 2.6246 862 2.6096 863 2.6163 864 2.5914 865 2.4822
```

```
866 2.5142 867 2.5387 868 2.6855 869 2.6055 870 2.5594
871 2.5299 872 2.5757 873 2.6837 874 2.4910 875 2.6280
876 2.5630 877 2.5316 878 2.5810 879 2.6983 880 2.6456
881 2.5005 882 2.4741 883 2.6105 884 2.5751 885 2.5389
886 2.6302 887 2.5293 888 2.6324 889 2.5820 890 2.4889
891 2.5897 892 2.6608 893 2.6072 894 2.5023 895 2.6234
896 2.6412 897 2.5181 898 2.5987 899 2.5359 900 2.6616
901 2.6380 902 2.5857 903 2.5584 904 2.5679 905 2.5452
906 2.6067 907 2.7003 908 2.6059 909 2.5032 910 2.6369
911 2.5847 912 2.5726 913 2.5914 914 2.5718 915 2.6408
916 2.5930 917 2.5925 918 2.5179 919 2.6700 920 2.5353
921 2.6665 922 2.6176 923 2.4578 924 2.5882 925 2.7009
926 2.6534 927 2.7135 928 2.5622 929 2.5848 930 2.5264
931 2.5954 932 2.5979 933 2.6942 934 2.5948 935 2.5379
936 2.5656 937 2.6412 938 2.5459 939 2.5708 940 2.6199
941 2.4470 942 2.5269 943 2.4246 944 2.6562 945 2.5668
946 2.6096 947 2.5676 948 2.6533 949 2.6158 950 2.5207
951 2.6321 952 2.5096 953 2.5925 954 2.6535 955 2.6583
956 2.5670 957 2.6334 958 2.6532 959 2.5287 960 2.5135
961 2.5883 962 2.6396 963 2.5662 964 2.5482 965 2.5591
966 2.5867 967 2.5747 968 2.5180 969 2.5271 970 2.7287
971 2.5975 972 2.6523 973 2.5830 974 2.6857 975 2.6935
976 2.5818 977 2.6885 978 2.5219 979 2.7201 980 2.5017
981 2.5105 982 2.6221 983 2.6324 984 2.6437 985 2.4997
986 2.5893 987 2.5385 988 2.5827 989 2.5777 990 2.5420
991 2.5807 992 2.4454 993 2.5244 994 2.5730 995 2.6106
996 2.4907 997 2.6830 998 2.6078 999 2.57341000 2.5762
```

List 10.2. The mean, standard deviation and 95% coverage interval for samples consisting of four values drawn from the population in list 10.1.

```
            60 samples of size 4 will now be taken:
      Row 1: Sample no., 4 sampled values with ident. nos.,
                     mean, st dev of mean

        Row 2: 95% C.I. using 1.96 x st dev of mean,
                     and verdict


 1    148 2.6269  433 2.5431  867 2.5387  922 2.6176   2.581563 0.023569
              2.535368  2.627757 so C.I. encloses m (2.581)

 2    674 2.5153  839 2.6002   13 2.5671  614 2.7109   2.598381 0.041396
              2.517246  2.679517 so C.I. encloses m (2.581)

 3    456 2.6919  874 2.4910  569 2.4508  106 2.5781   2.552936 0.053403
              2.448267  2.657605 so C.I. encloses m (2.581)

 4    393 2.5767  177 2.6938  575 2.5641  155 2.6387   2.618340 0.029976
              2.559586  2.677093 so C.I. encloses m (2.581)

 5     14 2.6177  679 2.4399  836 2.6282  818 2.5539   2.559921 0.043234
              2.475183  2.644659 so C.I. encloses m (2.581)

 6    467 2.6125  487 2.6493   95 2.6203  409 2.4893   2.592860 0.035411
              2.523454  2.662265 so C.I. encloses m (2.581)

 7     18 2.5569  467 2.6125  209 2.5500  345 2.5493   2.567178 0.015206
              2.537374  2.596982 so C.I. encloses m (2.581)

 8     30 2.4852  499 2.5311  574 2.6143  805 2.6429   2.568379 0.036472
              2.496893  2.639864 so C.I. encloses m (2.581)

 9    681 2.5661  419 2.5773  301 2.6138  520 2.6102   2.591867 0.011860
              2.568621  2.615113 so C.I. encloses m (2.581)

10    749 2.6492  971 2.5975  447 2.6526  685 2.6962   2.648887 0.020198
              2.609299  2.688475 so C.I. DOES NOT enclose m (2.581)

11    877 2.5316  682 2.5830  139 2.7399  324 2.5822   2.609166 0.045194
              2.520587  2.697746 so C.I. encloses m (2.581)

12    492 2.5698  989 2.5777  174 2.5406  970 2.7287   2.604211 0.042275
              2.521351  2.687071 so C.I. encloses m (2.581)

13    361 2.5320  794 2.5715  243 2.6210  271 2.5737   2.574556 0.018204
              2.538877  2.610235 so C.I. encloses m (2.581)

14    147 2.6410  397 2.5611  494 2.4806  560 2.4560   2.534667 0.041938
              2.452468  2.616866 so C.I. encloses m (2.581)

15    940 2.6199  779 2.5400  329 2.5213  130 2.6044   2.571421 0.024028
              2.524325  2.618517 so C.I. encloses m (2.581)

16    805 2.6429  939 2.5708  567 2.5893  659 2.6103   2.603323 0.015473
              2.572995  2.633651 so C.I. encloses m (2.581)

17    496 2.5534  801 2.5954  417 2.6966  149 2.6209   2.616592 0.030075
              2.557645  2.675539 so C.I. encloses m (2.581)
```

```
18    488 2.6057  129 2.6632  820 2.5766  667 2.6034    2.612235 0.018213
               2.576537  2.647933 so C.I. encloses m (2.581)

19    371 2.5832  870 2.5594  508 2.5605  946 2.6096    2.578179 0.011825
               2.555002  2.601357 so C.I. encloses m (2.581)

20    277 2.6443  714 2.5955  633 2.5366  135 2.6706    2.611744 0.029503
               2.553919  2.669569 so C.I. encloses m (2.581)

21     30 2.4852  580 2.6020  920 2.5353  504 2.7383    2.590193 0.054846
               2.482696  2.697691 so C.I. encloses m (2.581)

22    558 2.5864   36 2.4549  225 2.5931  825 2.5125    2.536730 0.032815
               2.472413  2.601047 so C.I. encloses m (2.581)

23     17 2.5796  946 2.6096  984 2.6437  202 2.5719    2.601192 0.016350
               2.569145  2.633239 so C.I. encloses m (2.581)

24    636 2.6552    1 2.6821   96 2.6520  597 2.5699    2.639761 0.024248
               2.592236  2.687286 so C.I. DOES NOT enclose m (2.581)

25    252 2.6834  697 2.5274  657 2.6354   71 2.5840    2.607551 0.033553
               2.541787  2.673315 so C.I. encloses m (2.581)

26      4 2.5869  945 2.5668  268 2.5990  470 2.5358    2.572114 0.013809
               2.545048  2.599181 so C.I. encloses m (2.581)

27    219 2.6257  506 2.6140  937 2.6412   53 2.6203    2.625280 0.005806
               2.613900  2.636660 so C.I. DOES NOT enclose m (2.581)

28    549 2.6604   60 2.5603  510 2.5651  597 2.5699    2.588903 0.023899
               2.542061  2.635746 so C.I. encloses m (2.581)

29    338 2.5184  260 2.4376   30 2.4852  478 2.4941    2.483811 0.016912
               2.450664  2.516958 so C.I. DOES NOT enclose m (2.581)

30    775 2.6130  838 2.6503  751 2.5991  798 2.4623    2.581171 0.041071
               2.500671  2.661671 so C.I. encloses m (2.581)

31    609 2.5646  229 2.5118  678 2.6790  154 2.6030    2.589636 0.035181
               2.520681  2.658590 so C.I. encloses m (2.581)

32    557 2.7288  503 2.6385   91 2.7514  555 2.6334    2.688020 0.030418
               2.628400  2.747640 so C.I. DOES NOT enclose m (2.581)

33      6 2.5632  496 2.5534    3 2.5799  854 2.5492    2.561452 0.006828
               2.548069  2.574835 so C.I. DOES NOT enclose m (2.581)

34    193 2.6273  452 2.4971  621 2.6332  646 2.6069    2.591123 0.031849
               2.528699  2.653546 so C.I. encloses m (2.581)

35    545 2.5529  497 2.7012  276 2.6371  772 2.5949    2.621511 0.031628
               2.559521  2.683501 so C.I. encloses m (2.581)

36    607 2.5363   47 2.6140  137 2.5535  139 2.7399    2.610932 0.046094
               2.520588  2.701276 so C.I. encloses m (2.581)

37    557 2.7288  939 2.5708  331 2.6195  515 2.5153    2.608573 0.045367
               2.519655  2.697491 so C.I. encloses m (2.581)

38    627 2.4806   68 2.6266  874 2.4910   32 2.7011    2.574829 0.053644
               2.469687  2.679971 so C.I. encloses m (2.581)

39    495 2.6605  140 2.6061  259 2.5950  376 2.4845    2.586519 0.036885
               2.514224  2.658814 so C.I. encloses m (2.581)

40    940 2.6199  333 2.6040  713 2.4589  593 2.4606    2.535844 0.044080
               2.449447  2.622241 so C.I. encloses m (2.581)

41    396 2.5263  535 2.5880  613 2.6295  964 2.5482    2.573005 0.022762
               2.528392  2.617619 so C.I. encloses m (2.581)

42    813 2.5310  993 2.5244  423 2.6167  299 2.6222    2.573583 0.026549
               2.521548  2.625619 so C.I. encloses m (2.581)

43    229 2.5118  206 2.5777  910 2.6369   70 2.6658    2.598056 0.034094
               2.531233  2.664879 so C.I. encloses m (2.581)

44     84 2.6124  404 2.5741  696 2.6744  892 2.6608    2.630431 0.023002
               2.585346  2.675516 so C.I. DOES NOT enclose m (2.581)

45    238 2.5321    1 2.6821  606 2.5711  828 2.6685    2.613446 0.036688
               2.541538  2.685354 so C.I. encloses m (2.581)

46     23 2.5861  452 2.4971  223 2.5865  867 2.5387    2.552105 0.021506
               2.509952  2.594257 so C.I. encloses m (2.581)

47    202 2.5719  186 2.5734  258 2.5968  373 2.6008    2.585712 0.007614
```

```
                    2.570787   2.600636  so C.I. encloses m (2.581)

48    914 2.5718  544 2.6424  316 2.6705  484 2.5386    2.605824 0.030560
                    2.545927   2.665721  so C.I. encloses m (2.581)

49    952 2.5096  108 2.5235  312 2.6541  782 2.5361    2.555821 0.033207
                    2.490735   2.620908  so C.I. encloses m (2.581)

50     48 2.5716  132 2.6149  205 2.7342  453 2.5545    2.618778 0.040502
                    2.539393   2.698163  so C.I. encloses m (2.581)

51    381 2.5754  259 2.5950  215 2.5902  985 2.4997    2.565065 0.022189
                    2.521574   2.608557  so C.I. encloses m (2.581)

52    120 2.6146  313 2.5297  253 2.6682  126 2.5178    2.582561 0.035751
                    2.512489   2.652632  so C.I. encloses m (2.581)

53    145 2.4313  180 2.6033  234 2.5616  334 2.5945    2.547662 0.039798
                    2.469657   2.625666  so C.I. encloses m (2.581)

54    465 2.6373  910 2.6369  694 2.5939  904 2.5679    2.609013 0.017071
                    2.575554   2.642471  so C.I. encloses m (2.581)

55     35 2.6743  805 2.6429  230 2.7241  936 2.5656    2.651702 0.033227
                    2.586578   2.716827  so C.I. DOES NOT enclose m (2.581)

56    224 2.6019   83 2.6305  560 2.4560  421 2.7304    2.604691 0.056701
                    2.493558   2.715824  so C.I. encloses m (2.581)

57    876 2.5630  149 2.6209  996 2.4907  805 2.6429    2.579369 0.034017
                    2.512696   2.646042  so C.I. encloses m (2.581)

58     34 2.5320  985 2.4997  618 2.5045  970 2.7287    2.566241 0.054632
                    2.459162   2.673320  so C.I. encloses m (2.581)

59    122 2.4949  145 2.4313  778 2.6047  402 2.6606    2.547864 0.051901
                    2.446138   2.649590  so C.I. encloses m (2.581)

60    246 2.5862  256 2.5221  386 2.5790  958 2.6532    2.585122 0.026842
                    2.532512   2.637733  so C.I. encloses m (2.581)

            Number of enclosures in 60 attempts:  52
                  or success rate:   87%
```

11

Case studies in measurement uncertainty

In this chapter we present four case studies based on typical undergraduate experiments, involving the determination of best estimates of measurands, standard uncertainties, expanded uncertainties and coverage intervals. For completeness, we include a brief description of each experiment. The equipment required is inexpensive or can usually be found in an undergraduate science laboratory. The account of each experiment contains data obtained in an actual experiment.

We have not included a detailed introduction to each experiment, nor have we indicated how each might be improved or 'finessed'. The account of each experiment is biased towards giving details of the data analysis such as the calculations of standard uncertainties and coverage intervals. A more detailed analysis would normally require consideration of the uncertainty in the calibration of instruments used. For many undergraduate experiments such information is not available, and therefore we have not included the contribution of the calibration uncertainty to the combined standard uncertainty. At the end of the account of each experiment we suggest practically based exercises related to the experiment.

11.1 Reporting measurement results

An account of an experiment, as presented in a formal report, may contain many sections with headings such as introduction, materials and methods, results, analysis and conclusion. With respect to the analysis of data, best estimates of particular quantities obtained through experiment and by other means should be communicated clearly, concisely, and in a manner that is useful to others. In particular, it is necessary to provide an account of the uncertainty components and how they were evaluated. Steps in the calculation of uncertainties should be sufficiently transparent that the calculation of (for example) a standard uncertainty can be verified by others. When calculating and reporting the best estimate of a quantity and uncertainty, we should do the following.

- Fully define the measurand. For example, if the electrical resistance of a metal wire is to be determined, the temperature at which the resistance is measured is an essential piece of information.
- State the best estimate of the measurand found by bringing together best estimates of the particular quantities that contribute to the calculation of the measurand. The unit of measurement of the measurand must be clearly stated.
- Describe the Type A and Type B evaluations of standard uncertainties that have been carried out. Show how these evaluations have been merged in the calculation of a combined standard uncertainty.
- Retain as many figures as possible in intermediate calculations, so that rounding errors do not accumulate. Once the expanded uncertainty has been determined, the best estimate of the measurand can be rounded to a 'sensible' number of significant figures. The analyses in this chapter were carried out with the aid of Excel. Excel retains 15 digits internally, and therefore it is assumed that rounding errors are negligible. We have chosen not to show all 15 digits in the intermediate calculations in this chapter.
- Use the Welch–Satterthwaite formula to determine the effective number of degrees of freedom, ν_{eff}. In order not to underestimate the coverage factor, k, ν_{eff} is rounded down to the nearest integer.
- Show how ν_{eff} has been used along with the chosen level of confidence to calculate the coverage factor, k.
- Quote the expanded uncertainty at the chosen level of confidence to two significant figures.
- State the coverage interval at the chosen level of confidence.

Advice regarding the calculation and expression of best estimates of measurands and their uncertainties is put into practice in the following case studies. Errors in replicate measurements are assumed uncorrelated, unless stated otherwise.

11.2 Determination of the coefficient of static friction for glass on glass

11.2.1 Purpose

The purpose of the experiment is to estimate the coefficient of static friction, μ_s, for glass on glass as well as the standard uncertainty in the estimate of μ_s and the coverage interval containing the true value of μ_s at the 95% level of confidence.

11.2.2 Background

The amount of force required to cause one body to slide over another depends on the nature of the two surfaces in contact. Consider a force applied to a body in a direction parallel to the surface on which the body rests. The force of static friction, F_s, which acts on the body (in the direction opposite to the applied force) increases in response to the applied force up to a maximum value, $F_{s,\max}$, at which point the

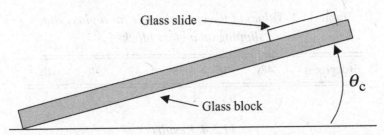

Figure 11.1. A glass slide on an inclined block of glass. When the angle of incline of the glass block equals the critical angle, θ_c, the glass slide begins to slip.

body will slip. $F_{s,max}$, is given by[1]

$$F_{s,max} = \mu_s N. \tag{11.1}$$

N is the force perpendicular to the surface exerted on the body by the surface with which it is in contact. μ_s is a dimensionless constant called the *coefficient of static friction*, which depends on the two surfaces in contact. In this experiment, μ_s is the measurand.

Typical values for μ_s are 0.1 for graphite on graphite and 1.5 for silver on silver.[2] Our goal in this experiment is to determine μ_s for glass sliding on glass.

One method of determining μ_s for two surfaces requires that a body made from, say, material A is placed upon another (usually larger) body made from material B. The bodies are tilted until they reach a critical angle, θ_c, at which point body A begins to slip. In this situation, it is possible to show that μ_s for the two surfaces in contact is related to θ_c by the equation

$$\mu_s = \tan \theta_c. \tag{11.2}$$

Through the determination of θ_c, we are able to find μ_s.

11.2.3 Method

A glass slide was placed on a block of glass[3] as shown in figure 11.1. The block of glass was inclined slowly until, at the critical angle, the glass slide began to slip. The critical angle was measured using a protractor with smallest scale interval of $1°$. The block and the slide were returned to their starting positions and the procedure repeated until six values of the critical angle had been obtained.

[1] See Halliday, Resnick and Walker (2004), chapter 6.
[2] See Serway and Faughn (2003), p. 101.
[3] Both the glass slide and the glass block were cleaned thoroughly with detergent then rinsed with water. The slide and the block were dried carefully.

Table 11.1. *Values of the critical angle for a glass slide slipping on a glass block*

x_i (degrees)	48	46	38	39	46	40

11.2.4 Results

Table 11.1 shows experimental values obtained for the critical angle.

11.2.5 Analysis

The best estimate of the critical angle, θ_c, may be written

$$\theta_c = X + Z, \tag{11.3}$$

where X is calculated by taking the mean of values obtained through repeat measurements. In the absence of systematic error, X is equal to θ_c. Z is the best estimate of the correction which accounts for the effect of systematic error.

Determination of X and the standard uncertainty in X
X is the mean of the values in table 11.1, so that $X = 42.83°$.

The standard uncertainty in X is based on a Type A evaluation of uncertainty. The standard deviation of the values in table 11.1, $s = 4.309°$. The number of degrees of freedom in the calculation of s is one fewer than the number of data, i.e., $\nu = 5$.

The standard uncertainty in X, $u(X)$, is given by

$$u(X) = \frac{s}{\sqrt{n}} = \frac{4.309°}{\sqrt{6}} = 1.759°. \tag{11.4}$$

The number of degrees of freedom associated with $u(X)$, which we write as ν_X, is the same as the number associated with s, i.e. $\nu_X = 5$.

Determination of Z and the standard uncertainty in Z
Several sources contribute to the best estimate of the correction, Z, including those due to the calibration error and resolution error. In this experiment we limit our determination of Z and the standard uncertainty in Z to consideration of the resolution error only, since we do not have information regarding other sources of error that may contribute to the correction term.

The correction due to resolution error alone could be either positive or negative. Since neither sign is favoured, we take the best estimate of the correction to be $Z = 0$. Using equation (11.3), it follows that $\theta_c = 42.83°$. The determination of the standard uncertainty in Z is not based on a statistical analysis, and therefore it is a Type B evaluation of uncertainty.

In this experiment, the resolution, δ, of the protractor[4] is $\delta = 1°$. The standard uncertainty in Z, $u(Z)$, is given by[5]

$$u(Z) = \frac{\delta}{\sqrt{12}},$$

i.e.

$$u(Z) = 0.289 \times 1° = 0.289°. \tag{11.5}$$

Since the uncertainty in $u(Z)$ is zero, the number of degrees of freedom associated with $u(Z)$ is taken to be very large, such that $\nu_Z \to \infty$ (see equation (9.18)).

The combined standard uncertainty
Inspection of equations (11.4) and (11.5) indicates that, in this experiment, $u(Z)$ is small compared with $u(X)$. It follows that $u(Z)$ could justifiably be neglected in the determination of the combined uncertainty. However, for completeness, we retain $u(Z)$ in the calculation of the combined standard uncertainty, $u(\theta_c)$, to give

$$u^2(\theta_c) = u^2(X) + u^2(Z) = (1.759°)^2 + (0.289°)^2 = (1.783°)^2.$$

The effective number of degrees of freedom, ν_{eff}
To calculate ν_{eff}, we use the Welch–Satterthwaite formula[6] which can be written in this situation as

$$\nu_{\text{eff}} = \frac{u^2(\theta_c)}{\dfrac{u^4(X)}{\nu_x} + \dfrac{u^4(Z)}{\nu_z}}. \tag{11.6}$$

As $\nu_Z \to \infty$, equation (11.6) simplifies to

$$\nu_{\text{eff}} = \frac{u^4(\theta_c)}{\dfrac{u^4(X)}{\nu_x}} = \frac{(1.783)^4}{\dfrac{(1.759)^4}{5}} = 5.3 \quad \text{(which truncates to 5).}$$

Calculation of the best estimate of the coefficient of static friction
μ_s is found by substituting the best estimate of the critical angle, $\theta_c = 42.83°$, into equation (11.2) to give

$$\mu_s = \tan(42.83) = 0.927. \tag{11.7}$$

---

[4] With care it is possible to estimate the angle reliably to the nearest 0.5°, in which case $\delta = 0.5°$. Here we use a slightly pessimistic estimate of the resolution.
[5] See section 5.5.
[6] See section 10.3 for a discussion of the Welch–Satterthwaite formula.

To find the standard uncertainty in μ_s, $\mu(\mu_s)$, we use the relationship

$$u^2(\mu_s) = \left(\frac{d\mu_s}{d\theta_c}u(\theta_c)\right)^2. \tag{11.8}$$

The derivative[7] $d\mu_s/d\theta_c$ is evaluated at $\theta_c = 42.83°$. For equation (11.8) to be valid, it is required that $u(\theta_c)$ be expressed in radians, i.e.

$$u(\theta_c) = 1.783° = 0.031\,11 \text{ rad}. \tag{11.9}$$

Differentiating equation (11.2) with respect to θ_c gives

$$\frac{d\mu_s}{d\theta_c} = \sec^2\theta_c. \tag{11.10}$$

Substituting $\theta_c = 42.83°$ into equation (11.10) gives

$$\frac{d\mu_s}{d\theta_c} = 1.859. \tag{11.11}$$

Substituting values for $u(\theta_c)$ and $d\mu_s/d\theta_c$ into equation (11.8) gives

$$u(\mu_s) = 0.0579.$$

The expanded uncertainty, $U(\mu_s)$, at the 95% level of confidence is given by

$$U(\mu_s) = ku(\mu_s), \tag{11.12}$$

where k is the coverage factor determined at a given level of confidence for a given number of degrees of freedom. By applying equation (11.6), we found the effective number of degrees of freedom to be $\nu_{\text{eff}} = 5$.

The coverage factor, k, at 95% level of confidence for five degrees of freedom is 2.57. Using equation (11.12), we find

$$U(\mu_s) = 2.57 \times 0.0579 = 0.149.$$

The coverage interval containing the true value of the coefficient of static friction at the 95% level of confidence is therefore (rounding the expanded uncertainty to two significant figures)

$$\mu_s \pm U(\mu_s) = 0.93 \pm 0.15.$$

11.2.6 Summary

The best estimate of the coefficient of static friction for glass on glass obtained in this experiment is $\mu_s = 0.93$.

---

[7] This derivative measures the sensitivity of μ_s to θ_c, and is an example of a sensitivity coefficient as described in section 7.1.1.

The standard uncertainty in the best estimate is $u(\mu_s) = 0.0579$. The effective number of degrees of freedom is $\nu_{eff} = 5$, giving a coverage factor of $k = 2.57$ for a 95% level of confidence.

The expanded uncertainty at the 95% level of confidence is $U(\mu_s) = 0.15$.

The coverage interval for the 95% level of confidence for the true value of the coefficient of static friction is 0.93 ± 0.15.

The value for the coefficient of static friction for glass on glass obtained in this experiment compares with the value of 0.94 published for glass on glass.[8]

Experimental exercise A

1. (a) Use a smooth flat piece of wood to act as an inclined plane. Determine the critical angle for a range of materials placed on the plane using the method described in section 11.2.3. Suggested materials are rubber, wood, glass and copper (or another metal).

   (b) Determine the best estimate of the coefficient of static friction and the coverage interval at the 95% level of confidence for each combination of materials in part (a) of this question. Compare your value for the coefficient of static friction of the material combinations with published values.

2. Investigate whether the coefficient of static friction is affected by surface smoothness. To do this, take one smooth glass slide and another glass slide that has been scratched using 'wet and dry' paper. Clean both carefully, then follow the method described in section 11.2.3. Through your analysis of the data, can you establish whether surface roughness is a factor that affects μ_s?

11.3 A crater-formation experiment

11.3.1 Purpose

The purpose of the experiment is to establish the relationship between the diameter, D, of a crater formed in sand and the kinetic energy, E, of a small ball striking the sand. In particular, the exponent, n, appearing in the equation, $D = cE^n$ is found using the experimental data. In addition, the standard uncertainty in n and the coverage interval at the 95% level of confidence are calculated.

11.3.2 Background

A crater is formed when a fast-moving object strikes the surface of, for example, a solid planet. By studying the relationship between the diameter of the crater and the kinetic energy of the impacting object, it is possible to discover which

---

[8] See Serway and Faughn (2003), p. 101.

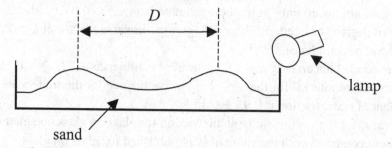

Figure 11.2. A crater formed when a steel ball strikes sand in a container.

energy-dissipating mechanism dominates (as examples, energy may be dissipated by deformation of material, ejection of material from the crater and the creation of seismic waves).

If the dominant process by which energy is dissipated is plastic deformation, then it is predicted that the diameter of the crater, D, should be given by[9]

$$D = cE^{1/3}. \qquad (11.13)$$

By contrast, if most of the incident kinetic energy is transferred to sand which is ejected from the crater, then the crater diameter is predicted to be related to the incident kinetic energy by the equation

$$D = cE^{1/4}. \qquad (11.14)$$

In equations (11.13) and (11.14), c is a constant.

11.3.3 Method

Steel balls of masses 8.35 g, 28.16 g and 66.76 g were dropped in turn from heights of between 25.5 cm and 150.0 cm into a container of 30 cm diameter filled with fine dry sand. The heights were chosen after preliminary measurements, which indicated that, owing to the relative insensitivity of the crater diameter to the kinetic energy of the ball, a wide range of kinetic energies should be employed in this experiment.

The sand was spread evenly to a depth of 10 cm. A small lamp was used to illuminate the sand in order to accentuate the contours of the crater. The diameter of the crater, D, as defined in figure 11.2, was measured using a plastic rule. The heights from which the balls were dropped were measured using a wooden metre rule. The smallest intervals marked on each rule were separated by 1 mm.

After measuring the diameter of the crater formed by the ball, the sand was shaken vigorously to ensure that the sand was not compacted. The sand was further

[9] See Amato and Williams (1998).

Table 11.2. *Values of crater diameter for various values of kinetic energy, E*

Mass, m(g)	Height, h (cm)	Kinetic energy, E(J)	Crater diameter, D (cm)
8.35	25.5	0.020 867	4.0, 4.0, 3.9
28.16	25.5	0.070 372	5.4, 5.3, 5.0
66.76	25.5	0.166 833	6.4, 6.4, 6.2
66.76	68.0	0.444 889	8.2, 7.8, 7.9
66.76	150.0	0.981 372	10.4, 10.0, 10.1

shaken (less vigorously) until the sand in the container was levelled. Three replicate measurements of crater diameter were made at each height for each ball used.

11.3.4 Results

Table 11.2 contains the values obtained for the diameter of the crater and the kinetic energy of the incident ball. The kinetic energies in table 11.2 were calculated assuming that all the potential energy possessed by a ball at height h is transformed into kinetic energy before impact. In this case we can write[10]

$$E = mgh, \qquad (11.15)$$

where m is the mass of the ball and g is the acceleration due to gravity. The acceleration due to gravity is taken as $9.80 \, \text{m/s}^2$, which is its value to three significant figures in Sydney, Australia, where the measurements were made.

11.3.5 Analysis

The relationship between the diameter of the crater, D, and the kinetic energy, E, of the impacting ball may be written

$$D = cE^n. \qquad (11.16)$$

Equation (11.16) can be fitted to data in table 11.2 using the technique of least-squares.

In applying least-squares we assume that

(a) the error is confined to the dependent variable (here the dependent variable is the diameter, D); and
(b) the size of scatter of the data about the line of best fit through the data should neither increase nor decrease over the range of the predictor variable.

---

[10] See Serway and Faughn (2003), p. 122.

To verify the validity of assumption (a) for this experiment, we compare the fractional uncertainty in E with the fractional uncertainty in D.

Uncertainty in the predictor variable, E
Through equation (11.15) we are aware that uncertainty in the best estimate of E depends on the uncertainties in the

- mass of the ball,
- acceleration due to gravity and
- height of fall of the ball.

Another source of uncertainty, which is not quantified here, and is assumed negligible, is due to the conversion of some of the kinetic energy of a falling ball into internal energy of the ball and the air due to air resistance (which causes the temperature of the ball and the air to increase slightly).

Since the resolution of the balance, δ, is $\delta = 0.01$ g, the standard uncertainty in the mass of the ball, $u(m)$, due to the limited resolution of the electronic balance is given by

$$u(m) = \delta/\sqrt{12} = 0.01 \times 0.189 \, \text{g}$$
$$= 0.002\,89 \, \text{g}.$$

The fractional standard uncertainty in the mass in this experiment, $u(m)/m$, for $m = 8.35$ g, is $(0.002\,89 \, \text{g})/8.35 \, \text{g} = 3.46 \times 10^{-4}$.

With respect to the acceleration due to gravity, g, we assume that g differs by no more than $0.01 \, \text{m/s}^2$ from the nominal value of $9.80 \, \text{m/s}^2$. Assuming that the distribution of g can be represented by a Gaussian distribution with standard uncertainty, $u(g) = 0.005 \, \text{m/s}^2$, the fractional standard uncertainty in g, $u(g)/g = (0.005 \, \text{m/s}^2)/(9.80 \, \text{m/s}^2) = 5.1 \times 10^{-4}$.

The uncertainty in the height measurement depends to an extent on the care taken when releasing the ball, in addition to how well the sand is levelled in the container. These uncertainties are likely to be greater than the uncertainty due to the limited resolution of the rule used to measure h, and so the uncertainty due to the limit of resolution of the rule is not considered in this analysis. We assume a Gaussian distribution for each measurement of height with a standard uncertainty, $u(h) = 1$ mm. It follows that $u(h)/h$ for $h = 255$ mm is $1 \, \text{mm}/255 \, \text{mm} \approx 0.004$.

The combined fractional uncertainty in the energy may be found by root-sum-squaring the fractional uncertainties in mass, acceleration due to gravity and height; this assumes that there is no correlation between the errors in the measurements of any of these quantities. This gives $u(E)/E \approx 0.0045$. Expressing the ratio as a percentage gives $u(E)/E \times 100\% = 0.45\%$.

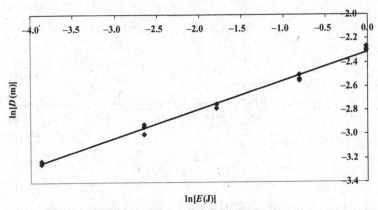

Figure 11.3. Variation of crater diameter with energy presented on a log–log scale.

Further calculations indicate that as the mass and height increase, so $u(E)/E$ decreases to about 0.001 (i.e. 0.1%) for $m = 66.76$ g and $h = 150$ cm.

Uncertainty in the dependent variable, D

Since three replicates of D have been made for each ball at each height, we may estimate the fractional standard uncertainty using a Type A evaluation of uncertainty at each energy. When the standard uncertainty, $u(D)$, in the mean value of D in table 11.2 is calculated for each value of kinetic energy, the fractional standard uncertainty, $u(D)/D$, is found to be in the range 0.0084 to 0.0230. Expressed as a percentage, this range is 0.84% to 2.3%.

As the fractional uncertainty in D is consistently greater than that in E, we proceed to analyse the data using unweighted linear least-squares in which we assume that error is confined to the dependent variable, D.

Least-squares analysis

To linearise equation (11.16) so that it is in the form $y = a + bx$, we take natural logarithms of both sides of equation (11.16), giving[11]

$$\ln D = \ln c + n \ln E \qquad (11.17)$$

$$y = a + b \; x \qquad (11.18)$$

i.e. $\ln c = a$, and $n = b$.

The natural logarithms of D and E in table 11.2 are calculated and are shown plotted in figure 11.3. An inspection of the graph of $\ln D$ versus $\ln E$ shown in

[11] We have chosen to use natural logarithms, though logarithms to any base would be equally valid.

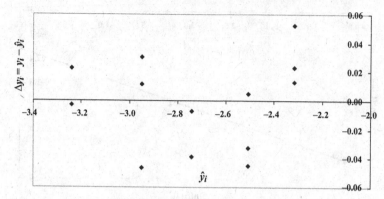

Figure 11.4. A plot of residuals indicating that the unweighted fit is valid.

figure 11.3 indicates that the linearisation has been successful. Included on the graph is the line of best fit obtained using least-squares.

Fitting equation (11.18) to data in table 11.2 using least-squares[12] gives a and b as

$$a = -2.3111, \qquad b = 0.2404.$$

It follows that $n = b = 0.2404$. Least-squares also gives the standard uncertainty in b, $u(b) = 0.005\,90$, so that $u(n) = 0.005\,90$.

In order to establish whether an unweighted fit to data is appropriate, the residuals, Δy_i, given by

$$\Delta y_i = y_i - \hat{y}_i \tag{11.19}$$

are plotted versus $\hat{y}_i$ as shown in figure 11.4. Here $y_i = \ln D_i$, where D_i is the ith value of the crater diameter as measured in the experiment. $\hat{y}_i = \ln \hat{D}_i$, where $\hat{D}_i$ is the ith value of D calculated at $E = E_i$ using the equation representing the line of best fit, as found using least-squares.

The plot in figure 11.4 shows no obvious trend in the residuals. This supports the assumption that the equation fitted to the data (i.e. equation (11.18)) is appropriate, since a mismatch between equation and data often causes a trend to appear in the residuals.[13]

Calculation of the expanded uncertainty in n at the 95% level of confidence
The expanded uncertainty, $U(n)$, is given by

$$U(n) = ku(n), \tag{11.20}$$

---

[12] The Excel spreadsheet by Microsoft was used to fit equation (11.18) to the data in table 11.2.
[13] See Devore (2003).

where k is the coverage factor determined at a given level of confidence for a given number of degrees of freedom.

Since a and b in equation (11.18) have been determined using 15 values, the number of degrees of freedom used in the determination of the coverage factor is that equal to that number of values $-$ 2, i.e. $\nu = 13$.

k at the 95% level of confidence for 13 degrees of freedom is equal to 2.16. Substituting this value for k into equation (11.20) gives the expanded uncertainty as

$$U(b) = 2.16 \times 0.0059 = 0.013 \qquad \text{(to two significant figures).}$$

The coverage interval for the 95% level of confidence for the true value of the exponent in equation (11.16) is therefore

$$n \pm U(n) = 0.240 \pm 0.013.$$

11.3.6 Summary

Using the data in this experiment, the best estimate of the exponent in equation (11.16) is $n = 0.240$.

The standard uncertainty in the best estimate is $u(n) = 0.0059$. The number of degrees of freedom is $\nu = 13$, giving a coverage factor of $k = 2.16$ for the 95% level of confidence.

The expanded uncertainty at the 95% level of confidence is $U(n) = 0.013$.

The coverage interval for the 95% level of confidence for the true value of the exponent is 0.240 ± 0.013.

The interval for n obtained through this experiment is consistent with the dominant energy-dissipation mechanism being due to ejection of material on impact of the ball with the sand, as suggested by equation (11.14).

Experimental exercise B
(a) Carry out this experiment using a coarse grade of sand.
(b) Determine the exponent, n, and the expanded uncertainty in the best estimate at the 95% level of confidence. Is the coverage interval for the 95% level of confidence obtained for n consistent with that expected for energy dissipation by ejection of material?

11.4 Determination of the density of steel

11.4.1 Purpose

The purpose of the experiment is to find the best estimate of the density, ρ, of a steel ball bearing at ambient temperature. The experiment requires the determination of the standard uncertainty in the best estimate of ρ and the expanded uncertainty at

Table 11.3. *Replicate values of the mass of the ball bearing*

Mass of steel ball bearing, x_{m_i} (g)	8.348	8.349	8.351	8.350	8.349	8.350	8.351	8.349

Table 11.4. *Replicate values of the diameter of the ball bearing*

Diameter of steel ball bearing, x_{d_i} (mm)	12.68	12.68	12.68	12.70	12.69	12.69

the 95% level of confidence. The value for the density is compared with published values for the density of steel.

11.4.2 Background

A fundamental property of any material is its density. If the mass of an object is M and the volume it occupies is V, then the average density of the material, ρ, is defined as

$$\rho = \frac{M}{V}. \tag{11.21}$$

11.4.3 Method

A steel ball bearing was weighed using a top-loading electronic balance with a resolution of 1 mg. Eight repeat measurements of the mass of the ball were made. Six repeat measurements were made of the diameter of the ball bearing using a micrometer. The smallest scale marks on the micrometer were separated by 0.01 mm. All measurements were made at $(23 \pm 1)\,^\circ$C.

11.4.4 Results

Table 11.3 contains the values obtained for the mass of the ball bearing obtained through repeat measurements. Table 11.4 contains values for the diameter of the same ball bearing measured at different positions around the ball.

11.4.5 Analysis

The volume, V, of a sphere of diameter D is written

$$V = \frac{\pi D^3}{6}. \tag{11.22}$$

This allows equation (11.21) to be written in terms of M and D, i.e.

$$\rho = \frac{M}{\dfrac{\pi D^3}{6}} = \frac{6M}{\pi D^3}. \qquad (11.23)$$

Best estimates of mass and diameter are combined to find the best estimate of the density of the ball bearing. To determine the standard uncertainty in the density, we need to determine the standard uncertainties both in the mass and in the diameter of the ball bearing taking into account Type A and Type B components.

Best estimate of mass and standard uncertainty in mass of the ball bearing
The best estimate, M, of the true mass is given by

$$M = X_m + Z_m. \qquad (11.24)$$

X_m is the mean of repeat measurements of the mass. $X_m = M$ in the absence of systematic errors. Z_m is a correction term introduced to account for the effect of systematic errors.

X_m is the mean of the value in table 11.3, i.e.

$$X_m = \frac{\sum\limits_{i=1}^{i=n} x_{m_i}}{n} = \frac{66.797}{8} = 8.3496 \text{ g}.$$

The estimate of the population standard deviation, s, of the values in table 11.3 is

$$s = 1.06 \times 10^{-3} \text{ g}.$$

The standard uncertainty, $u(X_m)$, in X_m is given by

$$u(X_m) = s/\sqrt{n} = 1.06 \times 10^{-3} \text{ g}/\sqrt{8} = 3.75 \times 10^{-4} \text{ g}.$$

The number of degrees of freedom, ν_{X_m}, associated with $u(X_m)$ is one fewer than the number of values, i.e. $\nu_{X_m} = 7$.

Determination of Z_m and the standard uncertainty in Z_m
The best estimate, Z_m, of the correction depends on several quantities such as calibration error and resolution error. In this experiment we limit our determination of Z_m and the standard uncertainty in Z_m to consideration of the resolution error only. The correction due to resolution error alone could be either positive or negative. Since neither sign is favoured, we take the best estimate of the correction to be $Z_m = 0$. It follows that the best estimate of the mass, M, is

$$M = X_m + Z_m = (8.3496 + 0) \text{ g} = 8.3496 \text{ g}. \qquad (11.25)$$

In this experiment, the resolution is $\delta = 1$ mg. The standard uncertainty in $u(Z_m)$, due to the limited resolution of the instrument, is[14]

$$u(Z_m) = \delta/\sqrt{12}$$
$$= 1 \times 0.289 \text{ mg} = 2.89 \times 10^{-4} \text{ g}. \tag{11.26}$$

Since the uncertainty in $u(Z_m)$ is zero, the number of degrees of freedom associated with $u(Z_m)$ is very large, i.e. $v_{z_m} \to \infty$.

The combined standard uncertainty in the mass, $u(M)$
The combined standard uncertainty in the mass, $u(M)$, is found using the equation

$$u^2(M) = u^2(X_m) + u^2(Z_m) = (3.75 \times 10^{-4} \text{ g})^2 + (2.89 \times 10^{-4} \text{ g})^2$$
$$= 2.24 \times 10^{-7} \text{ g}^2.$$

It follows that

$$u(M) = 4.73 \times 10^{-4} \text{ g}.$$

The effective number of degrees of freedom, v_{eff}, for the combined standard uncertainty in mass
To calculate v_{eff}, we use the Welch–Satterthwaite formula,[15] which can be written in this situation as

$$v_{\text{eff}} = \frac{u^4(M)}{\dfrac{u^4(X_m)}{v_{x_m}} + \dfrac{u^4(Z_m)}{v_{z_m}}} = \frac{5.05 \times 10^{-14}}{\left(\dfrac{1.98 \times 10^{-14}}{7}\right) + 0} = 17.8 \text{ (truncating to 17)}.$$

Uncertainty in the diameter of the ball bearing
The best estimate, D, of the diameter of the ball is given by

$$D = X_d + Z_d. \tag{11.27}$$

X_d is the diameter of the ball obtained by taking the mean of repeat measurements of the diameter. In the absence of systematic errors, $X_d = D$. The correction term introduced to account for the effect of systematic error is Z_d.

The mean diameter, X_d, is found using the data in table 11.4, i.e.

$$X_d = \frac{\sum x_{d_i}}{n} = \frac{76.13}{6} = 12.687 \text{ mm}.$$

[14] See section 5.5.
[15] See section 10.3 for a discussion of the Welch–Satterthwaite formula.

The estimate of the population standard deviation of the diameter values, s, found using the data in table 11.4 is

$$s = 8.16 \times 10^{-3} \, \text{mm}.$$

The standard uncertainty in the mean diameter, $u(X_d)$, of the ball is given by

$$u(X_d) = s/\sqrt{n} = 8.16 \times 10^{-3}/\sqrt{6} = 3.33 \times 10^{-3} \, \text{mm}.$$

The correction due to resolution error alone could be either positive or negative. Since neither sign is favoured, we take the best estimate of the correction to be $Z_d = 0$. This means that the best estimate of the diameter, D, is

$$D = X_d + Z_d = (12.687 + 0) \, \text{mm} = 12.687 \, \text{mm}. \tag{11.28}$$

In this experiment, the resolution of the micrometer is $\delta = 0.01$ mm. The standard uncertainty in $u(Z_d)$, due to the limited resolution of the instrument, is

$$u(Z_d) = \delta/\sqrt{12}$$
$$= 0.01 \times 0.189 \, \text{mm} = 2.89 \times 10^{-3} \, \text{mm}. \tag{11.29}$$

Since the uncertainty in $u(Z_d)$ is zero, the number of degrees of freedom associated with $u(Z_d)$ is very large, i.e. $\nu_{Z_d} \to \infty$.

The contribution to the combined uncertainty in the diameter due to other Type B components, such as that which may be expressed in a calibration certificate accompanying the micrometer, is not included in this analysis.

The combined standard uncertainty in diameter, $u(D)$
The combined standard uncertainty in the mass, $u(D)$, is found using the equation

$$u^2(D) = u^2(X_d) + u^2(Z_d) = (3.33 \times 10^{-3} \, \text{mm})^2 + (2.89 \times 10^{-3} \, \text{mm})^2$$
$$= 1.94 \times 10^{-5} \, \text{mm}^2,$$

i.e. $u(D) = 4.41 \times 10^{-3} \, \text{mm}.$

The effective number of degrees of freedom, ν_{eff}, for the combined standard uncertainty in the diameter
To calculate ν_{eff} we use

$$\nu_{\text{eff}} = \frac{u^4(D)}{\dfrac{u^4(X_d)}{\nu_{X_d}} + \dfrac{u^4(Z_d)}{\nu_{Z_d}}} = \frac{3.78 \times 10^{-10}}{\left(\dfrac{1.23 \times 10^{-10}}{5}\right) + 0} = 15.3 \; (\text{truncating to } 15).$$

Best estimate of density of the ball bearing

The best estimate of the density of the ball bearing is given by equation (11.23). Using the data in this experiment,

$$\rho = \frac{6 \times 8.3496}{\pi \times 12.687^3} = 7.810 \times 10^{-3} \, \text{g/mm}^3 \quad \text{(equivalent to } 7.810 \times 10^3 \, \text{kg/m}^3\text{)}.$$

Standard uncertainty in the density of the ball bearing

Regarding the errors in the measurement of diameter and mass as uncorrelated, the combined standard uncertainty in the density, $u(\rho)$, can be found using

$$u^2(\rho) = \left(\frac{\partial \rho}{\partial M} u(M) \right)^2 + \left(\frac{\partial \rho}{\partial D} u(D) \right)^2. \tag{11.30}$$

The partial derivatives in equation (11.30) are evaluated at the best estimate of mass and diameter, i.e. for $M = 8.3496 \, \text{g}$ and $D = 12.687 \, \text{mm}$, so that

$$\left(\frac{\partial \rho}{\partial M} \right) = \frac{6}{\pi D^3} = 9.353 \times 10^{-4} \, \text{mm}^{-3},$$

$$\left(\frac{\partial \rho}{\partial D} \right) = -\frac{18M}{\pi D^4} = -\frac{8.3496}{(1069.2)^2} = -1.847 \times 10^{-3} \, \text{g/mm}^4.$$

Substituting these partial derivatives into equation (11.30), together with the standard uncertainties in the mass and diameter, gives

$$u^2(\rho) = (9.353 \times 10^{-4} \times 4.73 \times 10^{-4})^2 + (-1.847 \times 10^{-3} \times 4.41 \times 10^{-3})^2$$
$$= 6.65 \times 10^{-11} \, (\text{g/mm}^3)^2.$$

It follows that $u(\rho) = 8.16 \times 10^{-6} \, \text{g/mm}^3$, which is equivalent to $8.16 \, \text{kg/m}^3$. Equation (11.30) can also be written

$$u^2(\rho) = u_1^2(\rho) + u_2^2(\rho), \tag{11.31}$$

where

$$u_1^2(\rho) = \left(\frac{\partial \rho}{\partial M} u(M) \right)^2, \qquad u_2^2(\rho) = \left(\frac{\partial \rho}{\partial D} u(D) \right)^2.$$

To find the 95% coverage interval of confidence for ρ, we need to determine the coverage factor, k. We begin by using the Welch–Satterthwaite formula to find the number of degrees of freedom, ν_{eff}, which can be written in this situation as

$$\nu_{\text{eff}} = \frac{u^4(\rho)}{\dfrac{u_1^4(\rho)}{\nu_1} + \dfrac{u_2^4(\rho)}{\nu_2}}. \tag{11.32}$$

Now

$$u^2(\rho) = 6.65 \times 10^{-11} \ (\text{g/mm}^3)^2$$

$$u_1^2(\rho) = \left(\frac{\partial \rho}{\partial M} u(M)\right)^2 = (9.353 \times 10^{-4} \times 4.73 \times 10^{-4})^2 = 1.96 \times 10^{-13} \ (\text{g/mm}^3)^2$$

$$u_2^2(\rho) = \left(\frac{\partial \rho}{\partial D} u(D)\right)^2 = (-1.847 \times 10^{-3} \times 4.41 \times 10^{-3})^2 = 6.63 \times 10^{-11} \ (\text{g/mm}^3)^2.$$

For the mass estimation, $v_1 = 17$. For the diameter estimation, $v_2 = 15$.

Substituting values into equation (11.32) gives

$$v_{\text{eff}} = \frac{(6.66 \times 10^{-11})^2}{\dfrac{(1.96 \times 10^{-13})^2}{17} + \dfrac{(6.63 \times 10^{-11})^2}{15}} = 15.1 \qquad \text{(truncating to 15).}$$

The coverage factor, k, for 15 degrees on freedom, and at the 95% level of confidence, is 2.13. It follows that the expanded uncertainty at the 95% level of confidence is given by

$$U(\rho) = 2.13 \times u(\rho) = 2.13 \times 8.16 \times 10^{-6} \ \text{g/mm}^3 = 1.74 \times 10^{-5} \ \text{g/mm}^3.$$

The coverage interval for the 95% level of confidence for the true value of the density is therefore

$$\rho \pm U(\rho) = (7.810 \pm 0.017) \times 10^{-3} \ \text{g/mm}^3.$$

11.4.6 Summary

The best estimate of the density of the ball bearing at 23 °C is $\rho = 7.810 \times 10^{-3} \ \text{g/mm}^3$.

The combined standard uncertainty in the best estimate of the density is $u(\rho) = 8.16 \times 10^{-6} \ \text{g/mm}^3$. The effective number of degrees of freedom is $v_{\text{eff}} = 15$, giving a coverage factor, $k = 2.13$, for a 95% level of confidence.

The expanded uncertainty at the 95% level of confidence, $U(\rho) = 1.74 \times 10^{-5} \ \text{g/mm}^3$.

The coverage interval for the 95% level of confidence for true value for the density, ρ, can be written

$$\rho \pm U(\rho) = (7.810 \pm 0.017) \times 10^{-3} \ \text{g/mm}^3.$$

This is equivalent to

$$(7.810 \pm 0.017) \times 10^3 \ \text{kg/m}^3.$$

We may compare the value obtained here for the density of the ball bearing with published values for the density of stainless-steels. While the largest component to stainless-steel alloys is iron, several other elements may also be present, such as nickel and chromium. The range of densities of stainless steel is normally in the range 7.73×10^{-3} g/mm^3 to 7.96×10^{-3} g/mm^3 as published by the company Goodfellow Metals.[16]

Experimental exercise C

An alternative way to determine the volume of a body that has density greater than that of water is to immerse it in water contained within a measuring cylinder. The volume of water displaced is equal to the volume of the body. Use this method to find the volume of an irregular-shaped solid metal object. Measure the mass of the object. Use your data to calculate the density of the object, the combined standard uncertainty in the density and the expanded uncertainty at the 95% level of confidence.

11.5 The rate of evaporation of water from an open container

11.5.1 Purpose

To determine the best estimate of the rate at which tap water in a shallow plastic container evaporates per unit area in air at room temperature and to find the expanded uncertainty in the best estimate at the 95% level of confidence.

11.5.2 Background

Evaporation is the process by which molecules escape from the surface of the liquid. The evaporation rate of water depends on many factors including the temperature and humidity of the atmosphere and the air velocity (Hisatake *et al.* 1995). Knowledge of evaporation rate over a range of conditions of humidity, temperature and rate of air flow can assist in accounting for certain changes in the Earth's climate. For example, measurements on evaporation rates have been used to support the theory of 'global dimming' (Roderick and Farquhar 2002).

11.5.3 Method

A round container was placed on a top-loading electronic balance, which has a resolution of 1 mg. The balance was zeroed using the tare facility. The container

---

[16] Details can be found at the Goodfellow web site at http://www.goodfellow.com.

Table 11.5. *Replicate values of the diameter of a container*

Diameter (cm)	4.00	3.95	3.95	4.00	3.95	3.90

Table 11.6. *Variation of mass of water in an open container as a function of time*

Time (s)	0	300	600	900	1200	1500	1800	2100	2400	2700	3000	3300	3600
Mass (g)	2.909	2.884	2.867	2.851	2.834	2.818	2.800	2.782	2.767	2.758	2.742	2.730	2.716

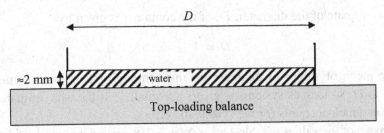

Figure 11.5. A schematic diagram showing a method for measuring the evaporation rate of water in an open container.

was then filled with tap water to an approximate depth of 2 mm as shown in figure 11.5. The balance was situated in a draught-free environment. The mass of water remaining in the open container was measured at time intervals of 300 s using a stopwatch capable of measuring time intervals to a resolution of 0.1 s. The temperature of the room was $(23 \pm 1)\,°C$. The relative humidity of the room was $(65 \pm 5)\%$. The diameter of the container was measured with callipers that have a resolution of 0.05 cm.

11.5.4 Results

Replicate values of the diameter of the container are given in table 11.5.

Table 11.6 contains 13 values obtained for the mass of water remaining in the container over a time interval of 3600 s.

11.5.5 Analysis

The expression for the evaporation rate, e
The best estimate of the evaporation rate per unit area, e, may be written

$$e = \frac{b}{A}, \tag{11.33}$$

where b is the best estimate of the evaporation rate and A is the area of the surface of the water exposed to the atmosphere. A can be written in terms of the diameter, D, of the vessel containing the water as

$$A = \frac{\pi D^2}{4}.$$ (11.34)

Equation (11.33) becomes

$$e = \frac{4b}{\pi D^2}.$$ (11.35)

The best estimate of the diameter, D, of the container is given by

$$D = X + Z.$$ (11.36)

X is the mean of values of diameter obtained through repeat measurements. X is equal to D, so long as systematic errors are small. Z is the best estimate of the correction which accounts for the systematic errors.

The mean of the values in table 11.5 is $X = 3.958$ cm and the standard deviation is $s = 0.037\,64$ cm.

The standard uncertainty of the mean of the values in table 11.5 is given by

$$u(X) = \frac{s}{\sqrt{n}} = \frac{0.037\,64}{\sqrt{6}} = 0.015\,37 \text{ cm.}$$

Since s was calculated using six values, the number of degrees of freedom is $6 - 1 = 5$, i.e. $v_X = 5$.

The best estimate of the correction, Z, depends on several quantities, such as calibration error and resolution error. In this experiment we limit our determination of Z and the standard uncertainty in Z to consideration of the resolution error only.

The limited resolution of the callipers of $\delta = 0.05$ cm introduces a Type B component of uncertainty. The correction due to resolution error alone could be either positive or negative. Since neither sign is favoured, we take the best estimate of the correction to be $Z = 0$. This means that the best estimate of the diameter, D, is

$$D = X + Z = (3.958 + 0) \text{ cm} = 3.958 \text{ cm.}$$ (11.37)

The standard uncertainty associated with the limited resolution is given by

$$u(Z) = \frac{0.05}{\sqrt{12}} \text{ cm} = 0.014\,43 \text{ cm.}$$

Since the uncertainty in $u(Z)$ is zero, the number of degrees of freedom associated with $u(Z)$ is taken to be very large, i.e. $v_Z \rightarrow \infty$.

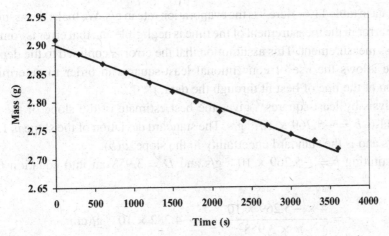

Figure 11.6. Variation of mass of water with time in an open container.

The combined standard uncertainty in the diameter, D
This is calculated using

$$u^2(D) = u^2(X) + u^2(Z) = (0.015\,37\,\text{cm})^2 + (0.014\,43\,\text{cm})^2 = 4.444 \times 10^{-4}\,\text{cm}^2,$$

so that $u(D) = 0.021\,08\,\text{cm}.$

The effective number of degrees of freedom, ν_{eff}, for the standard uncertainty in the diameter
To calculate ν_{eff} we use the Welch–Satterthwaite formula, which can be written in this case as

$$\nu_{\text{eff}} = \frac{u^4(D)}{\dfrac{u^4(X)}{\nu_X} + \dfrac{u^4(Z)}{\nu_Z}}. \tag{11.38}$$

As $\nu_Z \to \infty$, equation (11.38) simplifies to

$$\nu_{\text{eff}} = \frac{u^4(D)}{\dfrac{u^4(X)}{\nu_X}} = \frac{(0.021\,08)^4}{\dfrac{(0.015\,37)^4}{5}} = 17.7 \qquad \text{(truncating to 17)}.$$

Best estimate of slope of the mass-versus-time graph
Inspection of the graph of mass versus time, shown in figure 11.6, indicates that it is reasonable to fit an equation of the form

$$y = a + bx \tag{11.39}$$

to the data in table 11.6. Here b is the evaporation rate in g/s. We make the assumption that the error in the measurement of the time is negligible and that error is confined to the mass measurement. This assumption that the error is confined to the dependent variable allows the use of conventional least-squares in order to determine the equation of the line of best fit through the data.

Analysis by least-squares[17] gives the best estimate of the slope of the line in figure 11.6, $b = -5.269 \times 10^{-5}$ g/s. The standard deviation of the slope is 1.216×10^{-6} g/s and is the standard uncertainty in the slope, $u(b)$.

Substituting $b = -5.269 \times 10^{-5}$ g/s and $D = 3.958$ cm into equation (11.35) gives

$$e = \frac{4 \times -5.269 \times 10^{-5}}{\pi \times 3.958^2} = -4.282 \times 10^{-6} \text{ g/(cm}^2 \cdot \text{s)}.$$

The combined standard uncertainty in the evaporation rate

Regarding the errors in the slope of the line and diameter of the container as uncorrelated, the combined standard uncertainty in the evaporation rate, $u(e)$, can be found using

$$u^2(e) = \left(\frac{\partial e}{\partial b} u(b)\right)^2 + \left(\frac{\partial e}{\partial D} u(D)\right)^2. \tag{11.40}$$

The partial derivatives in equation (11.40) are evaluated at the best estimates, b and D.

Using equation (11.35),

$$\frac{\partial e}{\partial b} = \frac{4}{\pi D^2} = \frac{4}{\pi \times 3.958^2} = 0.081\,28 \text{ cm}^{-2},$$

$$\frac{\partial e}{\partial D} = \frac{-8b}{\pi D^3} = \frac{-8 \times -5.269 \times 10^{-5}}{\pi (3.958)^3} = 2.164 \times 10^{-6} \text{ g/(cm}^3 \cdot \text{s)}.$$

Substituting $\partial e/\partial b$ and $\partial e/\partial D$ into equation (11.40), together with $u(b)$ and $u(D)$, gives

$$u^2(e) = (0.081\,28 \times 1.216 \times 10^{-6})^2 + (2.164 \times 0.021\,08)^2$$
$$= 9.768 \times 10^{-15} + 2.081 \times 10^{-15}$$
$$= 1.185 \times 10^{-14} \text{ [g/(cm}^2 \cdot \text{s)]}^2.$$

It follows that

$$u(e) = 1.089 \times 10^{-7} \text{ g/(cm}^2 \cdot \text{s)}.$$

[17] The Excel spreadsheet by Microsoft was used to fit $y = a + bx$ to the data in table 11.6.

Equation (11.40) can be written

$$u^2(e) = u_1^2(e) + u_2^2(e),$$ (11.41)

where

$$u_1^2(e) = \left(\frac{\partial e}{\partial b} u(b)\right)^2, \qquad u_2^2(e) = \left(\frac{\partial e}{\partial D} u(D)\right)^2.$$ (11.42)

To find the 95% coverage interval for the evaporation rate per unit area, we find the effective number of degrees of freedom using the Welch–Satterthwaite formula. For this problem, ν_{eff} is given by

$$\nu_{eff} = \frac{u^4(e)}{\dfrac{u_1^4(e)}{\nu_1} + \dfrac{u_2^4(e)}{\nu_2}}.$$ (11.43)

We have already determined

$$u^2(e) = 1.185 \times 10^{-14} \ [\text{g/(cm}^2 \cdot \text{s})]^2,$$
$$u_1^2(e) = 9.768 \times 10^{-15} \ [\text{g/(cm}^2 \cdot \text{s})]^2, \qquad \nu_1 = 11,$$
$$u_2^2(e) = 2.081 \times 10^{-15} \ [\text{g/(cm}^2 \cdot \text{s})]^2, \qquad \nu_2 = 17.$$

It follows that

$$\nu_{eff} = \frac{(1.185 \times 10^{-14})^2}{\dfrac{(9.768 \times 10^{-15})^2}{11} + \dfrac{(2.081 \times 10^{-15})^2}{17}} = 15.7 \quad \text{(truncating to 15)}.$$

The coverage factor, k, and expanded uncertainty
The coverage factor, k, for the 95% level of confidence, when $\nu_{eff} = 15$, is $k = 2.13$.
The expanded uncertainty, $U(e)$, for the 95% level of confidence is given by

$$U(e) = ku(e) = 2.13 \times 1.089 \times 10^{-7} \ \text{g/(cm}^2 \cdot \text{s}) = 2.320 \times 10^{-7} \ \text{g/(cm}^2 \cdot \text{s}).$$

It follows that the coverage interval containing the true value of the evaporation rate per unit area at the 95% level of confidence is

$$e \pm U(e) = (-4.28 \pm 0.23) \times 10^{-6} \ \text{g/(cm}^2 \cdot \text{s}).$$

Further analysis
Close inspection of the line of best fit in figure 11.6 indicates that the scatter of the data about the line is not random, but exhibits a definite trend. This is further supported by the plot of residuals, $y_i - \hat{y}_i$ versus $\hat{y}_i$, where y_i is the measured mass remaining and $\hat{y}_i$ is the calculated mass remaining as found using the equation

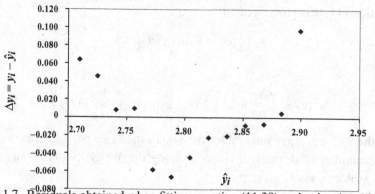

Figure 11.7. Residuals obtained when fitting equation (11.38) to the data in table 11.6.

$\hat{y}_i = a + bx_i$. The residuals are shown in figure 11.7. The trend from positive to negative then back to positive residuals is an indication that there is a model violation.[18] That is to say, the equation $y = a + bx$ is probably not optimum and another equation (perhaps a higher-order polynomial) should be considered.

11.5.6 Summary

For the conditions prevailing in this experiment, namely air temperature of $(23 \pm 1)\,°\mathrm{C}$, relative humidity $(65 \pm 5)\%$ and the container holding the water isolated from draughts, the best estimate of the evaporation rate for water per unit area in the container is $e = -4.28 \times 10^{-6}\,\mathrm{g/(cm^2 \cdot s)}$.

The standard uncertainty in the best estimate is $u(e) = 1.089 \times 10^{-9}\,\mathrm{g/(cm^2 \cdot s)}$. The effective number of degrees of freedom is $\nu_{\mathrm{eff}} = 15$, giving a coverage factor of $k = 2.13$ for a 95% level of confidence.

The expanded uncertainty at the 95% level of confidence for the true evaporation rate per unit area is therefore

$$U(e) = 2.3 \times 10^{-7}\,\mathrm{g/(cm^2 \cdot s)}.$$

The coverage interval for the 95% level of confidence for true evaporation rate is $(-4.28 \pm 0.23) \times 10^{-6}\,\mathrm{g/(cm^2 \cdot s)}$.

Experimental exercise D

To what extent does the evaporation rate of water per unit area depend on the surface area of the water? To investigate this, fill plastic containers of different areas with water to the same depth. Keeping other variables as constant as possible (such as

---

[18] See Kirkup (2002).

ambient temperature and local air flow), measure the evaporation rate per unit area as a function of area.

11.6 Review

In this chapter we have analysed data from experiments drawn from a range of topics often forming an element of an undergraduate laboratory programme. We have used methods described in the GUM to determine standard uncertainties, effective numbers of degrees of freedom and expanded uncertainties at the 95% level of confidence. In all the examples we have considered both Type A and Type B contributions to the total uncertainty. Type B uncertainties were based upon the limited resolution of the instruments used. In situations in which other uncertainty information is available, such as that found in a calibration report or certificate, that information should also be incorporated into the Type B uncertainty evaluation.

Appendix A

Solutions to exercises

Chapter 2

Exercise A

(a) $\mathrm{kg^{-1} \cdot m^{-3} \cdot s^4 \cdot A^2}$, (b) $\mathrm{kg \cdot s^{-3}}$, (c) $\mathrm{kg \cdot m^{-1} \cdot s^{-2}}$, (d) $\mathrm{kg \cdot m^2 \cdot s^{-2} \cdot K^{-1}}$, (e) $\mathrm{kg \cdot m^3 \cdot s^{-3} \cdot A^{-2}}$, (f) $\mathrm{kg \cdot s^{-3} \cdot A^{-2}}$, (g) $\mathrm{kg^{-1} \cdot s \cdot A}$, (h) $\mathrm{kg \cdot m^2 \cdot s^{-2} \cdot A^{-2}}$, (i) $\mathrm{kg^{-1} \cdot m^{-3} \cdot s^4 \cdot A^2}$, (j) $\mathrm{kg \cdot m^2 \cdot s^{-2}}$, (k) $\mathrm{kg \cdot m \cdot s^{-2} \cdot A^{-2}}$, (l) $\mathrm{kg \cdot s^{-3} \cdot K^{-4}}$

Exercise B

(a) 7.7 nC, (b) 52 pJ, (c) 7.834 kV, (d) 13 Mm/s, (e) 350 μPa · s

Exercise C

(a) 6.75×10^{-2} N, (b) 3×10^3 kg, (c) 1.6×10^{-19} C, (d) 7.55×10^{-1} V, (e) 3.5×10^{-3} kat, (f) 9.821×10^8 W

Exercise D

(a) 67.5×10^{-3} N, (b) 3×10^3 kg, (c) 160×10^{-21} C, (d) 755×10^{-3} V, (e) 3.5×10^{-3} kat, (f) 982.1×10^6 W

Exercise E

(a) 3.56 m, (b) 1.4×10^3 J/C or 1.4×10^3 V, (c) 11.85 g, (d) 3.24

Chapter 4

Exercise A

(1) variance $= 0.305$ mg^2, standard uncertainty $= 0.552$ mg
(2) variance $= 906.7$ nm^2, standard uncertainty $= 30.1$ nm

Exercise B

(a) 11.85 mg, 0.17 mg, (b) 423 nm, 12 nm.

Exercise C

(a) 5.557 N, (b) 0.078 N, (c) 0.023 N

Exercise D
(1) 38.8 cm/s, 2.0 cm/s
(2) 18.7 m/s, 1.3 m/s

Chapter 5

Exercise A
(1) 0.167, 0.180
(2) (b) 3, 11, 51

Exercise C
(1) 2.10×10^5 m, 58.09 m/s^2, 2.0272×10^6 (m/s)^2, 9.1×10^9 m^2
(2) 9.801 m/s^2, -3.400×10^{-6} s^{-2}

Exercise D
0.012 m/s^2, 3.1×10^{-7} s^{-2}

Chapter 6

Exercise A
1034.66 mbar, 0.08 mbar

Exercise B
(a) -4.22 mV
(b) 3.24×10^{-5} V
(c) 3.6×10^{-5} V

Exercise C
Add 0.0154 V to the value indicated by the DMM.

Exercise D
$+10.5$ μg/g in the reported mass

Exercise E
52.8 °C

Chapter 7

Exercise A
(1) 46.5 Hz, 5.2 Hz
(2) 12.73, 0.16
(3) (a) expressions for $\partial v/\partial T$ and $\partial v/\partial \mu$: $1/(2\sqrt{\mu T})$ and $-\frac{1}{2}\sqrt{T/\mu^3}$, respectively
    (b) 49.32 m/s, 0.57 m/s
(4) (a) $q^2/(p+q)^2$, $p^2/(p+q)^2$, (b) 9.66 cm, 0.31 cm

Exercise B

(a) $c_1 = \dfrac{2x_1 x_2}{x_3}$, $c_2 = \dfrac{x_1^2}{x_3}$, $c_3 = -\dfrac{x_1^2 x_2}{x_3^2}$

(b) $c_1 = 2^{-3/2}(x_1 x_2)^{-1/2}$, $c_2 = -(x_1/(2x_2)^3)^{1/2}$

(c) $c_1 = \exp x_2$, $c_2 = x_1 \exp x_2$,

(d) $c_1 = \dfrac{\cos x_1}{\sin x_2}$, $c_2 = \dfrac{-\sin x_1 \cos x_2}{\sin^2 x_2}$

Exercise C

1064.6 Ω, 1.240 mA, 0.033 mA

Exercise D

322.5 nm, 2.8 nm

Exercise E

(a) 30.53 cm, 5.44 cm
(b) 0.10 cm, 0.068 cm
(c) 5.611
(d) 0.072

Chapter 8

Exercise A

0.246, 0.0547

Exercise B

(1) 5.8, 1.83, 3.36
(2) (b) $\frac{1}{2}$, (c) 0.3125

Exercise C

0.14 °C, 0.058 mL, 2.9 pF, 0.0029 s

Exercise D

(1) (b) 0, 0.845, (c) 0.0313, (d) 0, 0.345
(2) (a) 0.500, 0.289, (b) 0.500, 0.204

Chapter 9

Exercise A

(1) 50
(2) 0.25

Chapter 10

Exercise A

(1) (a) 9.075 L/mg, -1.53, (b) 0.0979 L/mg, 0.8157, (c) 0.272 L/mg, 2.27, (d) 8.80 L/mg to 9.35 L/mg, -3.79 to $+0.74$
(2) 0.237 μV/V (yr)$^{-1}$ to 0.267 μV/V (yr)$^{-1}$

Exercise B

(a) 712.5 cm³

(b) 8.75 cm³

(c) 6.4

(d) 2.45 for six degrees of freedom

(e) 691.1 cm³ to 733.9 cm³

Exercise C

(1) (a) (i) 13.8 μV, (ii) 14 degrees of freedom, (iii) 29.5 μV

(b) (i) 12.0 μV, (ii) 11 degrees of freedom, (iii) 26.4 μV

(2) (a) (i) 9.5 μV, (ii) 9 degrees of freedom, (iii) 21.4 μV

(b) (i) 6.7 μV, (ii) 19 degrees of freedom, (iii) 14.0 μV

The solution to this problem indicates that, with no systematic error, the expanded uncertainty is reduced by more than 30% if the number of readings is doubled; but with the systematic error, the reduction is only by slightly more than 10%.

Appendix B

95% Coverage factors, k as a function of the number of degrees of freedom, v

Degrees of freedom, v	Coverage factor, k
2	4.30
3	3.18
4	2.78
5	2.57
6	2.45
7	2.36
8	2.31
9	2.26
10	2.23
11	2.20
12	2.18
13	2.16
14	2.14
15	2.13
16	2.12
17	2.11
18	2.10
19	2.09
20	2.09
25	2.06
30	2.04
40	2.02
50	2.01
100	1.98
Infinite	1.96

Appendix C

Further discussion following from the Welch–Satterthwaite formula

The effective number of degrees of freedom associated with the uncertainty of a measurand can never exceed the sum of the degrees of freedom associated with the uncertainties of the inputs.

This is a consequence of the Welch–Satterthwaite formula discussed in section 10.3. For n inputs $x_1, x_2, \ldots, x_n$ with standard uncertainties $u(x_1), u(x_2), \ldots, u(x_n)$, sensitivity coefficients $c_1, c_2, \ldots, c_n$ and degrees of freedom $\nu_1, \nu_2, \ldots, \nu_n$, the Welch–Satterthwaite formula states that

$$\frac{\left[c_1^2 u^2(x_1) + c_2^2 u^2(x_2) + \cdots + c_n^2 u^2(x_n)\right]^2}{\nu_{\text{eff}}} = \frac{c_1^4 u^4(x_1)}{\nu_1} + \frac{c_2^4 u^4(x_2)}{\nu_2} + \cdots + \frac{c_n^4 u^4(x_n)}{\nu_n}. \quad (1)$$

From (1) it follows that

$$\nu_{\text{eff}} \leq \nu_1 + \nu_2 + \cdots + \nu_n. \quad (2)$$

This may be shown by algebraic manipulation of (1), but a demonstration in terms of electric circuits may be of interest.[1] For convenience of illustration figure C.1 shows the particular case of five inputs, $n = 5$, but the following argument applies in an obvious way to the general case of n inputs.

In figure C.1(a) the batteries have voltages $c_1^2 u^2(x_1), c_2^2 u^2(x_2), \ldots, c_n^2 u^2(x_n)$, and are connected across resistances $\nu_1, \nu_2, \ldots, \nu_n$. When a battery of voltage V is connected across a resistance R, the power dissipated in the resistance is V^2/R. So the total power dissipation P_1 in all the resistances is, in figure C.1(a) (for n batteries and resistors),

$$P_1 = \frac{c_1^4 u^4(x_1)}{\nu_1} + \frac{c_2^4 u^4(x_2)}{\nu_2} + \cdots + \frac{c_n^4 u^4(x_n)}{\nu_n}. \quad (3)$$

In figure C.1(b), all internal links are removed. The batteries are now all in series, connected across all the resistances in series, and so the total power dissipation P_2 in all the resistances is

$$P_2 = \frac{\left[c_1^2 u^2(x_1) + c_2^2 u^2(x_2) + \cdots + c_n^2 u^2(x_n)\right]^2}{\nu_1 + \nu_2 + \cdots + \nu_n}. \quad (4)$$

Since conducting material has been removed in going from figure C.1(a) to figure C.1(b), and the circuit in figure C.1(a) consists only of constant voltage sources and linear resistances,

---

[1] The algebraic manipulation and the electric-circuit demonstration are both described in Frenkel (2003).

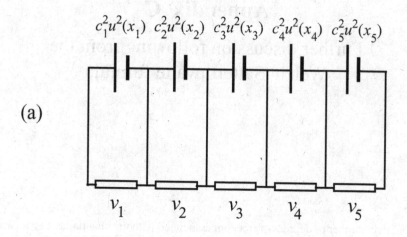

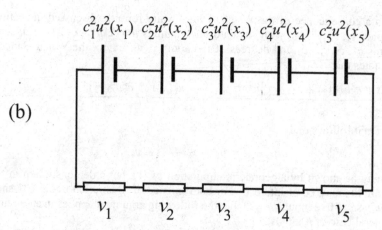

Figure C.1. Electrical analogue to the Welch–Satterthwaite formula.

P_2 must be less than P_1. (If all the battery voltages are in the same mutual ratios as their corresponding resistances, so that $c_1^2u^2(x_1)/[c_2^2u^2(x_2)] = v_1/v_2$, etc. then P_2 is equal to P_1, and the equality in (2) holds. The internal links would then not have carried any current anyway in figure C.1(a). If there were just two batteries and resistances obeying the ratio – and therefore only one internal link – this particular case would be recognised as essentially a Wheatstone-bridge circuit in balance, with no current in the link.) The fact that P_2 is less than P_1 may be checked as plausible by considering simple circuits of batteries and resistors.[2] So if $P_2 < P_1$, then

$$\frac{\left[c_1^2u^2(x_1) + c_2^2u^2(x_2) + \cdots + c_n^2u^2(x_n)\right]^2}{v_1 + v_2 + \cdots + v_n} < \frac{c_1^4u^4(x_1)}{v_1} + \frac{c_2^4u^4(x_2)}{v_2} + \cdots + \frac{c_n^4u^4(x_n)}{v_n}. \quad (5)$$

---

[2] The fact that $P_2 < P_1$ may be proven rigorously in a more general context of a linearly conducting medium. See, for example, Ferraro (1958), chapter 12.

The right-hand side of (5) is also the right-hand side of (1). So (5) and (1) together give

$$\frac{\left[c_1^2 u^2(x_1) + c_2^2 u^2(x_2) + \cdots c_n^2 u^2(x_n)\right]^2}{v_1 + v_2 + \cdots + v_n} < \frac{\left[c_1^2 u^2(x_1) + c_2^2 u^2(x_2) + \cdots c_n^2 u^2(x_n)\right]^2}{v_{\text{eff}}}, \quad (6)$$

or, on cancelling out the equal numerators in (6),

$$v_{\text{eff}} < v_1 + v_2 + \cdots + v_n. \quad (7)$$

References

(Papers related to topics discussed in chapter 1 are listed at the end of that chapter.)

Alder, K. (2002), *The Measure of All Things: The Seven-Year Odyssey and Hidden Error that Transformed the World*, London, Abacus.

Allan, D. W. (1987), 'Should the classical variance be used as a basic measure in standards metrology?', *IEEE Trans. Instrum. Meas.*, **IM-36**, 646–654.

Amato, J. C. and Williams, R. E. (1998), 'Crater formation in the laboratory', *American J. Phys.*, **66**, 141–143.

Ballico, M. (2000), 'Limitations of the Welch–Satterthwaite approximation for measurement uncertainty calculations', *Metrologia*, **37**, 61–69.

Balsamo, A., Mana, G. and Pennecchi, F. (2005), 'On the best fit of a line to uncertain observation pairs', *Metrologia*, **42**, 376–382.

Barnard, F. A. P. (1872), *The Metric System of Weights and Measures*, New York, van Nostrand.

Bendat, J. S. and Piersol, A. G. (2000), *Random Data: Analysis and Measurement Procedures*, New York, John Wiley and Sons.

Bentley, R. E. (2005), *Uncertainty in Measurement: The ISO Guide*, Technology Transfer Series Monograph No. 1, Sydney, National Measurement Institute of Australia.

Bevington, P. R. and Robinson D. K. (2002), *Data Reduction and Error Analysis for the Physical Sciences*, 3rd edn, New York, McGraw-Hill.

Blaisdell, E. A. (1998), *Statistics in Practice*, 2nd edn, Fort Worth, Saunders College Publishing.

Cantor, R. and Koelle, D. (2004), 'Practical DC SQUIDs: configuration and performance', in *The SQUID Handbook*, vol. 1, ed. J. Clarke and A. Braginski, Weinheim, Wiley-VCH Verlag GmbH and Co.

Clothier, W. K., Sloggett, G. J., Bairnsfather, H., Currey, M. F. and Benjamin, D. J. (1989), 'A determination of the volt', *Metrologia*, **26**, 9–46.

Cox, M. G. and Harris, P. M. (2004), *Uncertainty Evaluation*, Software Support for Metrology Best Practice Guide No. 6, National Physical Laboratory, UK. Also at http://www.npl.co.uk/ssfm/download/documents/ssfmbpg6.pdf.

Davis, R. S. (2002), 'The SI unit of mass', *Metrologia*, **40**, 299–305.

Decker, J. E. and Pekelsky, J. R. (1996), *Uncertainty of Gauge Block Calibration by Mechanical Comparison: A Worked Example. Case 1: Gauges of Like Material*, Document 39998, National Research Council of Canada.

Devore, J. L. (2004), *Probability and Statistics for Engineering and the Sciences*, 6th edn, Belmont, California, Brookes/Cole.

Draper, N. and Smith, H. (1981), *Applied Regression Analysis*, 2nd edn, New York, John Wiley and Sons.

Elster, C. (2000), 'Evaluation of measurement uncertainty in the presence of combined random and analogue-to-digital conversion errors', *Meas. Sci. Technol.*, **11**, 1359–1363.

Ferraro, V. C. A. (1958), *Electromagnetic Theory*, London, Athlone Press, University of London.

Flyvbjerg, H. and Petersen, H. G. (1989), 'Error estimates on averages of correlated data', *J. Chem. Phys.*, **91**, 461–466.

Frenkel, R. B. (2003), *Statistical Background to the ISO 'Guide to Expression of Uncertainty in Measurement'*, Technology Transfer Series Monograph No. 2, Sydney, National Measurement Laboratory (now National Measurement Institute of Australia).

Frenkel, R. B. and Kirkup, L. (2005), 'Monte Carlo-based estimation of uncertainty owing to limited resolution of digital instruments', *Metrologia*, **42**, L27–L30.

Hall, B. D. and Willink, R. (2001), 'Does Welch–Satterthwaite make a good uncertainty estimate', *Metrologia*, **38**, 9–15.

Halliday, D., Resnick, R. and Walker, J. (2004), *Fundamentals of Physics*, 7th edn, New York, Wiley.

Harris, I. A. and Warner, F. L. (1981), 'Re-examination of mismatch measurements when measuring microwave power and attenuation', *IEE Proc. Part II – Microwaves, Optics and Antennas*, **128**, part H, 35–41.

Hisatake, K., Fukuda, M., Kimura, J., Maeda, M. and Fukuda, Y. (1995), 'Experimental and theoretical study of evaporation of water in a vessel', *J. Appl. Phys.*, **77**, 6664–6674.

ISO (1993), *Guide to the Expression of Uncertainty in Measurement*, Geneva, International Organisation for Standardisation (corrected and reprinted 1995).

Kacker, R. and Jones, A. (2003), 'On use of Bayesian statistics to make the Guide to the Expression of Uncertainty in Measurement consistent', *Metrologia*, **40**, 235–248.

Kendall, M. G. and Stuart, A. (1969), *The Advanced Theory of Statistics*, vol. 1, 3rd edn, London, Charles Griffin and Co.

Kirkup, L. (2002), *Data Analysis with Excel: An Introduction for Physical Scientists*, Cambridge, Cambridge University Press.

Klein, H. A. (1989), *The Science of Measurement*, New York, Dover Publications.

Kutner, M. J., Nachtsheim, C. J. and Neter, J. (2004), *Applied Linear Regression Models*, 4th edn, New York, McGraw-Hill/Irwin.

Limpert, E., Stahel, W. A. and Abbt, M. (2001), 'Lognormal distributions across the sciences: keys and clues', *Bioscience*, 51, 341–352.

Lira, I. H. and Wöger, W. (1997), 'The evaluation of standard uncertainty in the presence of limited resolution of indicating devices', *Meas. Sci. Technol.*, **8**, 441–443.

Macdonald, J. R. and Thompson, W. J. (1992), 'Least-squares fitting when both variables are subject to error: pitfalls and possibilities', *American J. Phys.*, **60**, 66–73.

Malakoff, D. (1999), "Bayes offers a 'new' way to make sense of numbers", *Science*, **286**, 1460–1464.

Mills, I. M., Mohr, P. J., Quinn, T. J., Taylor, B. N. and Williams, E. R. (2005), 'Redefinition of the kilogram: a decision whose time has come', *Metrologia*, **42**, 71–80.

Nicholas, J. V. and White, D. R. (2001), *Traceable Temperatures: An Introduction to Temperature Measurement and Calibration*, Chichester, Wiley.

Pritchard, B. J. (1997), 'Production of 1 ohm resistors at the National Measurement Laboratory', *Proceedings of Metrology Society of Australia*, 149–151.

Quinn, T. J., Speake, C. C., Richman, S. J., Davis, R. S. and Picard, A. (2001), 'A new determination of *G* using two methods, *Phys. Rev. Lett.*, **87**, 111101-1–111101-4.

Roderick, M. L. and Farquhar, G. D. (2002), 'The cause of decreased pan evaporation over the past 50 years', *Science*, **298**, 1410–1411.

Rose-Innes, A. C. and Rhoderick, E. H. (1977), *Introduction to Superconductivity*, 2nd edn, Oxford, Pergamon Press.

Seber, G. A. F. (1977), *Linear Regression Analysis*, New York, John Wiley and Sons.

Serway, R. A. and Faughn, J. S. (2003), *College Physics*, 6th edn, Pacific Grove, California, Brookes/Cole.

Vinal, G. W. (1950), *Primary Batteries*, New York, John Wiley and Sons.

Wilks, S. S. (1962), *Mathematical Statistics*, London, John Wiley and Sons.

Witt, T. J. (2000), 'Testing for correlation in measurements', in *Proceedings of Advanced Mathematical and Computational Tools in Metrology*', Singapore, World Scientific, pp. 273–288; also in (2003) *Using the Allan Variance in DC Electrical Measurements*, PTB Colloquium, Braunschweig, Physikalisch-Technische Bundesanstalt.

Witt, T. J. and Reymann, D. (2000), 'Using power spectra and Allan variances to characterise the noise of Zener-diode voltage standards', *IEE Proc. Sci. Meas. Technol.*, **147**, 177–182.

Young, H. D. and Freedman, R. A. (2003), *University Physics with Modern Physics*, 11th edn, Reading, Massachusetts, Addison Wesley.

Index

Page numbers followed by f refer to footnotes.

Printed in the United States
By Bookmasters